Essential Discrete Mathematics for Computer Science

Todd Feil
Department of Mathematics and Computer Science
Denison University

Joan Krone
Department of Mathematics and Computer Science
Denison University

PEARSON EDUCATION, INC., Upper Saddle River, NJ 07458

Library of Congress Cataloging-in-Publication Data

Feil, Todd.
Essential discrete mathematics for computer science / Todd Feil and Joan Krone.
p. cm.
Includes bibliographical references and index.
ISBN: 0-13-018661-9
1. Computer science--Mathematics. I. Krone, Joan. II. Title

QA76.9.M35 F44 2003 2002030816
004'.01'51--dc21

Editor-in-Chief: *Sally Yagan*
Acquisition Editor: *George Lobell*
Vice-President/Director of Production and Manufacturing: *David W. Riccardi*
Executive Managing Editor: *Kathleen Schiaparelli*
Senior Managing Editor: *Linda Mihatov Behrens*
Assistant Managing Editor: *Bayani Mendoza de Leon*
Production Editor: *Jeanne Audino*
Manufacturing Buyer: *Michael Bell*
Manufacturing Manager: *Trudy Pisciotti*
Marketing Manager: *Halee Dinsey*
Marketing Assistant: *Rachel Beckman*
Art Director: *Jayne Conte*
Cover Designer: *Bruce Kenselaar*
Cover Photo: *Wallhanging made of wool and cotton by Benita Koch-Otte, 1923/24.*
Photo from Bauhaus-Archive, Berlin, from Bauhaus Weavings, Berlin 1998, Nr. 148.

Pearson Education, Inc.
Upper Saddle River, New Jersey 07458

Printed in the United States of America

10 9 8 7 6 5 4 3 2 1

ISBN 0-13-018661-9

Pearson Education LTD., *London*
Pearson Education Australia PTY, Limited, *Sydney*
Pearson Education Singapore, Pte. Ltd
Pearson Education North Asia Ltd, *Hong Kong*
Pearson Education Canada, Ltd., *Toronto*
Pearson Educación de Mexico, S.A. de C.V.
Pearson Education -- Japan, *Tokyo*
Pearson Education Malaysia, Pte. Ltd

Contents

Preface

If you are an instructor, why should you choose this textbook for your students? If you are a student, why should you read this text? The material included in this text provides an introduction to discrete mathematics and is intended for first year students so that their later courses in mathematics and/or computer science can be covered in more depth than they could be without this foundational background. The text is not intended to be a comprehensive collection of discrete mathematics topics, but rather it ties selected topics to concepts in computer science and it includes programming problems along with written exercises. Unlike the large, comprehensive texts, this one can be covered in a semester. For computer science students, there are programming exercises. For math students without an interest in programming, there are plenty of exercises of different levels to challenge them.

This text evolved over a 10-year period from notes for our second semester freshman course for computer science students. This course has included about two-thirds mathematics and about one-third programming. Our students have found immediate benefits in their next course, Data Structures and Algorithms Analysis, as well as all other upper level courses. You will find the style focused on the chosen topics; we make no attempt at a complete coverage of those concepts. We chose the topics with two goals in mind: to lay a strong mathematical foundation and to show that mathematics has immediate application in computer science.

There is little, if any, controversy over whether or not computer science students should study mathematics. The resounding consensus is that mathematics is critical to the study and practice of computer science. It is not so easy to gain agreement among academicians and practitioners as to exactly what areas of mathematics should be studied, how rigorous the presentation should be, and at what points in the curriculum these ideas should be introduced.

Having been involved in the education of computer science students and having been responsible for teaching students who have taken a variety of mathematics courses, it is our belief that it is wise to include some fundamental mathematics in the first-year computer science curriculum. We believe that there are enough topics for which students can see immediate applications that it is worthwhile to make those topics a part of the CS1 or CS2 course. This is not to say that students would not need or benefit from other courses in mathematics in addition to what they learn at this point. Rather, we believe that students will enjoy and get more from later mathematics courses because they have some background in basic ideas.

This book is not intended to be "the" math course for computer science students. It is intended to help students understand the importance of mathematics and see its relevance in a variety of applications. Indeed, most students will take some sort of discrete mathematics course later in their careers. The most immediate application for students is in analyzing algorithms, something they will start doing in earnest in their next course or two. To understand not only standard arithmetic algorithms but also important algorithms in cryptology, students must understand modular arithmetic and basic number theory. Concerns arise later that require a foundation in mathematics.

Precision of expression is the key to carrying out the tasks of both program specification and program correctness, and mathematics provides the foundation for this precision. Mathematics teaches us to be exact in what we say and how we think. It gives us the capability to express our ideas in such a way as to avoid being misunderstood. The study of mathematics in general, regardless of specific content, promotes precision of expression and attention to detail in reasoning. However, we have chosen particular mathematical structures that have direct applications in computer science, hence addressing two goals. First, we concern ourselves with the task of helping students develop reasoning skills and exactness of expression. Our second goal is to provide the fundamental mathematics necessary for computer scientists.

Many exercises are included at the end of each chapter. Some suggestions for programming problems have been included. Most are easily embellished or altered to meet the needs of the course. Some exercises and programming problems have been starred. These indicate more challenging problems.

> ✓ Throughout the text you will find questions displayed like this. These are usually straightforward questions to be done as the student reads the text to check if the material is understood.

Many people aided in the creation of this text. We'd like to thank first our students who, over many semesters, pointed out errors in the text (typographic and other) and offered suggestions about exercises. We'd like to mention particularly Rohit Bansal and Tony Fressola in this regard.

The editorial staff at Prentice-Hall has been particularly helpful: Patricia Daly, Jeanne Audino and George Lobell. The original manuscript has come a long way thanks to them.

And finally, we'd like to thank our spouses, Robin and Gil, for their encouragement and support.

Any errors and typos are, of course, our responsibility. We would like to hear from you if you find any. Please email us with any errors you find or comments you have about the text.

Todd Feil
feil@denison.edu

Joan Krone
krone@denison.edu

Chapter 0
Introduction to Proofs

Throughout this text, we have provided proofs for many of the theorems presented and have assigned some proofs in the exercises. Proofs are an important part of mathematics. But figuring out how to start a proof and how to proceed is especially difficult for someone new to this activity. The purpose of this chapter is to present a couple of proof techniques used in the text, give some simple yet illustrative examples, and supply some justification on why these techniques work. There are many other techniques used besides the ones we include here; this is only a starting point. Please note that one of the most important techniques, induction, is delayed until later; we have devoted Chapter 4 to this powerful method.

A proof is a convincing argument. Outside of mathematics, what constitutes a proof differs from the high standards set in mathematics. For example, if you wanted to prove that you climbed Mount Everest, you might supply photos of yourself taken on the summit or a letter from someone attesting to the fact or even the results of a lie detector test. A skeptic might protest that the photos and letters could be faked and the lie-detector fooled and therefore the evidence offers no proof at all. The courtroom is a source of many such "proofs." Bear in mind that mathematicians are the most skeptical of people, at least when it pertains to mathematics.

One of the things that separates mathematics from other intellectual endeavors is the preciseness of the claims made. This requires a certain formality in the language, and the proofs of these claims usually require the same formality. But it's true that not all proofs are formal. A goal of a proof is to communicate to someone the argument being made. (That someone may be yourself.) So the level of detail offered depends on the audience. If you are a professional mathematician and you are communicating a proof to another mathematician who is an expert in the field, a proof might be a few sentences or paragraphs. The details that to an expert are well-known or easily worked out are skipped. This "handwaving" is common. An expert needs the main ideas of the proof. A less sophisticated audience needs more details to see the connections between steps in the proof. Indeed a very naive audience would be so ignorant of the subject that a great deal of effort would be necessary to make the terms intelligible. The real test here would be to keep the audience's interest maintained over the days and weeks (or longer) required.

What differs in the level of sophistication in various versions of a proof is the size of the steps used. The audience must believe that each step is justified. A mathematician believes that every theorem can be reduced to a series of formal statements, starting from a system of axioms, whose steps would be so simple that they would be easy, even trivial, to justify. However, the argument would be so long that the essence of the proof would be lost.

At the other extreme, a proof given in conversation between experts would involve huge steps, leaving a great deal of "detail" for the listener to justify. These steps, you could argue, give only the essence of the proof.

Fortunately, over the years, mathematicians have developed a style that is a blend of natural language and formalism that has evolved into a balance of preciseness and readability so you believe it would be possible to reduce the proof to a series of strictly formal statements (even though this would be a Herculean task). Your ability to read and produce proofs partly involves learning this style. We start with a short introduction to logic.

Propositional Logic

A proposition is a statement that has an associated truth

value. That is, a proposition is either true or false. Examples are: *$3 + 5 = 7$. Wood comes from trees. There are 53 states.* (The first and last propositions are false and the second one is true.) Not all statements are propositions: *Red is a pretty color. This sentence is false.* The former statement is a matter of opinion; it might be true for some people and false for others. The latter statement is a logical paradox; assuming it is true would lead you to conclude it is false, and assuming it is false would lead you to conclude it is true. In either case, you have a contradiction. We are not concerned with statements of either of these two types.

> ✓ State three propositions. State three sentences that are not propositions and tell why.

Propositions may occur in more complicated forms: *If it rains, we will cancel the picnic. His mother is a doctor, and his father is a painter. Either you will complete your work or you will not get credit.* Several propositions may be considered together: *If a function is not continuous, then it is not differentiable. If a function is not integrable, then it is not continuous. f is not differentiable. f is not continuous.*

To assign a truth value to a proposition it is necessary that some context or interpretation be established. For example *Today is Monday* must be assigned a truth value based on the use of some calendar system. But once truth values have been assigned to simple propositions, we can progress to more complicated statements. We can reason about these compound and complex propositions once we know the truth value for their simple components.

To reason about propositions, we use a syntax for representing propositions and a set of rules for manipulating these representations. These rules are called the *propositional calculus.*

To talk about propositions it is convenient to introduce some standard symbols. We'll use the symbols "t" and "f" for truth symbols (standing for "true" and "false") and $\wedge$, $\vee$, and $\neg$ for connectives. (These are called *and, or*, and *not* respectively. Sometimes *and* is called conjunction and *or* is called disjunction.) Propositional variables will be indicated by p, q, $r, \ldots$ and parentheses will be used in the usual way for punctuation. Propositional variables may have a value of "t" or "f."

The connectives can be explained by the following truth table:

p	q	$\neg p$	$p \wedge q$	$p \vee q$
t	t	f	t	t
t	f	f	f	t
f	t	t	f	t
f	f	t	f	f

Another possible connective is one called "xor," suggesting exclusive or. The meaning of "xor" is that one or other of two propositions must be true, but not both, in order for p xor q to be true.

> ✓ Make a truth table for xor.

We illustrate these with two simple propositions: Let p be the proposition "The sun is shining" and let q be the proposition "It is raining." Then $\neg p$ is "The sun is not shining", $p \wedge q$ is " The sun is shining and it is raining" and $p \vee q$ is "The sun is shining or it is raining." Of course whether each of these more complex complicated propositions is true or false depends on the truth value of p and q.

> ✓ Use the propositions for p and q as given in the previous paragraph. Look outside and determine if p is true or false. Do likewise for q. Now determine the truth of $p \wedge q$, $p \vee q$, $\neg p$, and $\neg q$.

There are two important special propositions: the proposition that always has the value of true, which we'll write as T, and the proposition that always has the value of false, which we'll write as F. (Note the difference between t and f (lowercase), which are possible values for propositions, and T and F, which are propositions that are always true or false, respectively. It is important to note this difference.) The symbols T and F are sometimes referred to as truth symbols.

As with arithmetic, we have rules of precedence for evaluating a logical expression: Whatever is in parenthesis should be evaluated first; $\neg$ has the highest priority, followed by $\wedge$, and then $\vee$.

For example, $\neg p \wedge q$ is different in meaning from $\neg(p \wedge q)$. In the first expression, $\neg$ is applied only to the proposition p, but in the second expression, the negation applies to the entire conjunction.

To give meaning to these expressions, we note that for any logical expression there is a unique truth table. Indeed, a logical expression is completely described by its truth table. Two expressions are said to be equivalent if and only if they have the same truth table. For example, suppose we want to show the distributivity of $\wedge$ over $\vee$. That is, $p \wedge (q \vee r) \equiv (p \wedge q) \vee (p \wedge r)$. To show this equivalence we simply give the truth table for the expression on the left-hand side and the truth table for the expression on the right-hand side and observe that they are the same. We combine these two truth tables into one, also giving some intermediate values. The last two columns are the ones we wish to compare.

p	q	r	$p \wedge q$	$p \wedge r$	$q \vee r$	$p \wedge (q \vee r)$	$(p \wedge q) \vee (p \wedge r)$
t	t	t	t	t	t	t	t
t	t	f	t	f	t	t	t
t	f	t	f	t	t	t	t
t	f	f	f	f	f	f	f
f	t	t	f	f	t	f	f
f	t	f	f	f	t	f	t
f	f	t	f	f	t	f	f
f	f	f	f	f	f	f	f

Note that the truth values for the expressions $p \wedge (q \vee r)$ and $(p \wedge q) \wedge (p \vee r)$ are the last two columns of the truth table and these two columns are identical. Thus the two expressions are equivalent and we write $p \wedge (q \vee r) \equiv (p \wedge q) \wedge (p \vee r)$.

It is important to observe that all the reasoning about the complicated propositions can take place without knowing what they say. The truth table indicates the truth values for every possible truth assignment to the simple statements p and q whatever they are.

> ✓ If p is the statement "Bob is over six feet tall", q is the statement "Sally is hungry", and r is the statement "Tom has blue hair", write the two statements $p \wedge (q \vee r)$ and $(p \wedge q) \vee (p \wedge r)$ as English sentences.

The following are some important facts about propositional logic. You are asked to prove these assertions in the Exercises.

- $\wedge$ and $\vee$ are commutative; that is, $p \wedge q \equiv q \wedge p$ and similarly for $\vee$.
- $\wedge$ and $\vee$ are associative; that is, $p \wedge (q \wedge r) \equiv (p \wedge q) \wedge r)$ and similarly for $\vee$.
- $\vee$ distributes over $\wedge$; that is, $p \vee (q \wedge r) \equiv (p \vee q) \wedge (p \vee r)$. (We've just shown that $\wedge$ distributes over $\vee$; that is, $p \wedge (q \vee r) \equiv (p \wedge q) \vee (p \wedge r)$.)
- T acts as an identity for $\wedge$ and F acts as an identity for $\vee$. (That is, $T \wedge p = p$ and $F \vee p = p$ for every proposition p.)
- $p \wedge (\neg p) = F$ and $p \vee (\neg p) = T$.

Implication

There is one more fundamental connective: *implication*, which uses the symbol $\Rightarrow$. We write $p \Rightarrow q$ and say "p implies q." This connective is important because most theorems will be of the form $p \Rightarrow q$. As with the connectives *and*, *or*, and *not*, the meaning of *implies* follows the usual English usage, as is given by the following truth table.

p	q	$p \Rightarrow q$
t	t	t
t	f	f
f	t	t
f	f	t

You should not have any difficulty understanding this truth table, which we'll take as the definition of *implies*, except perhaps with the third line, which says that *false* $\Rightarrow$ *true* is a true statement. We will postpone discussing why this is so for a few paragraphs. First let's look at some slightly more complicated expressions involving implies.

When evaluating an expression, $\Rightarrow$ has the lowest precedence; that is, we evaluate it last. For example, consider the expression $\neg q \Rightarrow p \wedge q$. By the rules of precedence, we evaluate $\neg$, then $\wedge$ and finally $\Rightarrow$. (You may wish to parenthesize this expression to be sure to make the precedence clear: $(\neg q) \Rightarrow (p \wedge q)$.) So the truth table for $\neg q \Rightarrow p \wedge q$ would be

p	q	$\neg q$	$p \wedge q$	$\neg q \Rightarrow p \wedge q$
t	t	f	t	t
t	f	t	f	f
f	t	f	f	t
f	f	t	f	f

Notice that this is equivalent to the expression q. Now consider the expression $\neg q \Rightarrow \neg p$, whose truth table is

p	q	$\neg q$	$\neg p$	$\neg q \Rightarrow \neg p$
t	t	f	f	t
t	f	t	f	f
f	t	f	t	t
f	f	t	t	t

Notice that this expression is equivalent to $p \Rightarrow q$. This is an important observation. We call $\neg q \Rightarrow \neg p$ the *contrapositive* of $p \Rightarrow q$, and we will discuss this more later.

Most theorems in mathematics are statements of the form $A \Rightarrow B$, where A and B are propositions. We would say "A implies B" or "If A then B." A is called the *hypothesis* and B is called the *conclusion.* We wish to prove the proposition $A \Rightarrow B$ is always true. For example, consider the statement "If n is an even integer then n^2 is also." Here, A (the hypothesis) is the proposition "n is an even integer" and B (the conclusion) is the proposition "n^2 is an even integer." Thus this statement is of the form $A \Rightarrow B$.

When we say the theorem $A \Rightarrow B$ is true, we mean it is always true. For this to be the case, obviously we need to show that if A is true, it follows that B must be true also—this is where the work is involved in proving the theorem. But what if A is false? In this case we don't care what the truth value of B. But we still want the theorem $A \Rightarrow B$ to be true. This is a justification why false $\Rightarrow$ true and false $\Rightarrow$ false are both ture statements.

Thus to show that $A \Rightarrow B$ is true, we need only establish that if we assume A is true it must follow that B is true. The argument that establishes this is called the proof of the theorem. There are many techniques for proving a theorem. We will show examples of the most common techniques.

In the assertion $A \Rightarrow B$ (A implies B), A is called the *hypothesis* and B is called the *conclusion.* You wish to prove this proposition $A \Rightarrow B$ is always true. For example, consider the statement *If n is an even integer, then n^2 is also.* Here, the hypothesis is the proposition *n is an even integer*, and the conclusion is the proposition *n^2 is an even integer.*

Direct Proof

A direct proof of the assertion $A \Rightarrow B$ is to assume the hypothesis (proposition A) is true and then argue that the conclusion (proposition B) must then be true.

Let's try a direct proof of the statement "If n is an even integer then n^2 is also." We assume that n is an even integer.

That means that $n = 2k$, for some integer k. This is typical of the first step in a direct proof. You ask the question "What does it mean to say 'n is an even integer'?" Or you ask "What is an immediate consequence of 'n is an even integer'?" An answer to the first type of question is usually the definition of the proposition in question. (That's what we've done in our case.) The answer to the second type of question is usually the result of another theorem you already know to be true. (Imagine you knew the theorem "every even integer is a floozlewhopper." Then a first step might be to say that "n is a floozlewhopper.")

Now that we know $n = 2k$, we wish to show n^2 is even. How could that be done? If we could show that $n^2 = 2m$, for some integer m, that would show that n^2 is even. Notice that we've started at the beginning (proposition A) and seen what we can easily derive from it. Then we go to the end (proposition B) and see what we need to show in order to conclude B is true. This forward-backward thinking continues until we close the gap between beginning and end and are able to supply an argument to justify a jump from our series of propositions derived from A to our series of propositions leading to B. In many cases this could be quite difficult, of course. Our example is fairly simple, though. For if $n = 2k$, then $n^2 = 4k^2 = 2(2k^2)$. So, letting $m = 2k^2$, we have written $n^2 = 2m$ and so n^2 is even.

Let's review this technique: We wish to show $A \Rightarrow B$. We first find a statement A_1 so that $A \Rightarrow A_1$. Maybe we can continue getting statements $A_2, A_3, \ldots A_j$ so that $A \Rightarrow A_1 \Rightarrow A_2 \Rightarrow \cdots \Rightarrow A_j$. We go as far as is reasonable. Then we find B_1 so that $B_1 \Rightarrow B$. Again maybe we find more statements so that $B_i \Rightarrow \cdots \Rightarrow B_1 \Rightarrow B$. Eventually we'll add to the A_j's and B_i's until we can show $A_j \Rightarrow B_i$. We then have that $A \Rightarrow B$. (This follows since $A \Rightarrow C$ and $C \Rightarrow B$ implies that $A \Rightarrow B$.) This completes the proof.

Usually when writing a proof of this type no forward-backward thinking is evident; the proof appears to be a straightforward series of implications. Here's a more "polished" version of our proof.

Theorem. *If n is an even integer, then so is n^2.*

Proof: Since n is even, there is an integer k such that $n = 2k$. Then $n^2 = 4k^2 = 2(2k^2)$ and so n^2 is even.

The Contrapositive

Another important proof technique uses the contrapositive. The basis for this technique is to see that $A \Rightarrow B$ is equivalent to its contrapositive $\neg B \Rightarrow \neg A$, which we've seen. Instead of proving the original implication, we prove the contrapositive, which is equivalent. So we assume that $\neg B$ is true and then prove $\neg A$ must also be true. This would establish the truth of $\neg B \Rightarrow \neg A$, which is equivalent to $A \Rightarrow B$.

Let's use the contrapositive to prove the statement "If n^2 is even, then so is n." The contrapositive to this statement is "If n is not even, then n^2 is not even." But since an integer that is not even must be odd, we can restate the contrapositive as "If n is odd, then n^2 is odd." This statement is fairly straightforward to prove directly. Recall that if an integer is odd then the integer is of the form $2k+1$ for some integer k. Thus, we start our proof by assuming that n is odd and so there is an integer k so that $n = 2k + 1$. Then $n^2 = (2k + 1)^2 = 4k^2 + 4k + 1 = 2(2k^2 + 2k) + 1$ and so n^2 is odd, as desired.

Proof by Contradiction

Another technique closely related to the contrapositive is proof by contradiction. When using this technique to prove $A \Rightarrow B$, let A be true (as usual) but also assume that B is false. From these assumptions we conclude that another statement C is true, but in fact we know that C is false. Or we show that C is false when we know that C is in fact true. (We say, here, that we've contradicted C.) Thus our assumption (that B is false) must have been in error. Hence B must be true. That is $A \Rightarrow B$, as we wish.

The difficulty with this technique, in contrast to the direct approach and proof by using the contrapositive, is that you do not know what the statement C is that you are going to contradict. Note that any statement C whose truth we know will serve as a statement to contradict. This prevents us from using the forward and backward arguments when developing a proof. Nonetheless it is a very important technique. Proof by contradiction is a good choice to prove a theorem of the form "every such-and-such is *not* a so-and-so." In this case, we assume we have a such-and-such that is a so-and-so and proceed to show this leads to a contradiction. We'll give two examples to illustrate this proof technique.

Our first example is to show that $\sqrt{2}$ is irrational. Since *irrational* means "not rational" we see that this is a good candidate for proof by contradiction. We start by assuming that $\sqrt{2}$ *is* rational. A number is rational if it can be written as a quotient of two integers, a/b. We may assume that a/b is in lowest terms; otherwise we simply reduce a/b.

But if $\sqrt{2} = a/b$, then $b\sqrt{2} = a$, and so by squaring both sides we see that $2b^2 = a^2$. Thus a^2 is even. But then a is even also. (We've shown this in a previous example.) So $a = 2k$, for some integer k. Thus $a^2 = 4k^2$ and so $2b^2 = 4k^2$. But then $b^2 = 2k^2$ and so b^2 is even, and hence b is even also. But then both a and b are even and this contradicts the fact that a/b is in lowest terms. Our assumption, that $\sqrt{2}$ is rational, must be false and so $\sqrt{2}$ is irrational.

Notice that the statement we contradicted in the last proof was that a/b was in lowest terms. You should agree that when starting the proof it was not obvious that this was the statement to be contradicted.

Another example of using proof by contradiction is to prove that there are an infinite number of primes. Again, *infinite* means "not finite," so you see this is a good candidate for proof by contradiction. We assume that there are only a finite number of primes. Suppose there are n primes; we'll index them $p_1, p_2, \ldots, p_n$. Now consider $m = p_1 p_2 \cdots p_n + 1$. By the Fundamental Theorem of Arithmetic (see the Chapter 5) m is either prime or divisible by a prime. But clearly m is not divisible by any of the p_i since dividing m by any of them results in a remainder of 1. This contradicts the Fundamental Theorem of Arithmetic and so our assumption is false; we conclude that there is an infinite number of primes.

If And Only If

Finally, many theorems we see will be of the form A is equivalent to B, which is sometimes written $A \Leftrightarrow B$. This means that both $A \Rightarrow B$ and $B \Rightarrow A$. So when proving $A \Leftrightarrow B$, you must prove two implications. When stating a theorem of this type, we frequently say "A if and only if B." Or we abbreviate by writing "A iff B." Earlier, we showed that n even implies that n^2 is even and we also showed that n^2 is even implies that n is even. Thus we have shown that n is even if and only if n^2 is even. (That is, n is even iff n^2 is even.)

Exercises

1. Prove that $\wedge$ and $\vee$ are commutative; that is, $p \wedge q \equiv q \wedge p$ and similarly for $\vee$.
2. Prove that $\wedge$ and $\vee$ are associative; that is, $p \wedge (q \wedge r) \equiv (p \wedge q) \wedge r$ and similarly for $\vee$.
3. Prove that $\vee$ distributes over $\wedge$. That is, show that $p \vee (q \wedge r) \equiv (p \vee q) \wedge (p \vee r)$.
4. Prove that T acts as an identity for $\wedge$ and that F acts as an identity for $\vee$. (That is, $T \wedge p \equiv p$ and $F \vee p \equiv p$ for every proposition p.)
5. Prove that $p \wedge (\neg p) \equiv F$ and $p \vee (\neg p) \equiv T$.
6. Show that if n is odd then so is n^2. (An odd integer is one of the form $2k + 1$.)

7. Show, by contrapositive, that n^2 is odd implies n is odd.
8. Show that $\sqrt{3}$ is irrational. (Use a technique similar to our proof that $\sqrt{2}$ is irrational.)
9. Show that $\sqrt[3]{2}$ is irrational.
10. Show that n^3 is odd iff n is odd.
11. Let p_i be the ith smallest prime. (So $p_1 = 2$, $p_2 = 3$, and so on.) Note that $2 + 1 = 3$ is prime, $2 \cdot 3 + 1 = 7$ is prime, and $2 \cdot 3 \cdot 5 + 1 = 31$ is prime. But if you calculate $p_1 p_2 \cdots p_n + 1$ for various n, you don't always get a prime. Find the smallest n where this fails.
12. Prove, by contradiction that the set of integers is infinite.
13. Suppose the average of three different integers is 20. Prove that at least one of the three must be greater than 20. (What technique will you use here?)
14. Suppose the average of four different integers is 20. Prove that at least one of the four must be greater than 21.
15. Show that at least one of the digits $1, 2, \ldots, 9$ must appear infinitely often in the decimal expansion of π.
16. Prove by contradiction that if there are n (> 2) people at a party, then at least 2 people have the same number of friends at the party.

Chapter 1
Sets

The language of set theory is used to express ideas in all of mathematics. Most questions about the nature of mathematics can be reduced to questions in set theory. The concepts and symbols provide tools for reasoning in most areas of mathematics. A formal approach to axiomatic set theory, such as the Zermelo-Fraenkel system, requires a significant amount of time and background. It is not our intent to pursue a formal approach here. Rather, we want to become familiar with basic concepts and symbols of set theory to the extent that we will be able to use them in expressing the mathematics throughout the rest of the text. The concepts and notation of set theory provide students with important skills for reading and expressing ideas.

What Are Sets?

Sets are very simple mathematical objects. For an object S to be a set, it is necessary only that we can tell for any item x whether or not x is in S. If x is in S, we say *x is an element of S* and write $x \in S$. If x is not in S, we write $x \notin S$. One way of expressing a set is to list the elements of a set between the symbols $\{$ and $\}$, such as $\{2, 5, 6, 8\}$ or $\{2, 3, 4, \ldots\}$. Another way is to use the so-called set-builder notation, where we list the properties the items in the set should have, such as $\{x \in \mathbb{R} : x^2 < 100x + 10\}$. We will not always be able to do this, however. Indeed, we will see later that the use of set-builder notation may lead to descriptions for entities that are not sets!

For any particular discussion, items for membership consideration come from what we call a universal set. For example, we might restrict our consideration to the set of integers, which would then be our universal set. Frequently, the universal set will be understood, but sometimes we will need to state explicitly our universal set. Computer scientists frequently deal with more than one universal set in a program—this is analogous to using different data types. For programming, these various universal sets need to be identified explicitly.

> ✓ There may be many choices for a universal set. If a computer program is written to sort numbers, what are some appropriate universal sets for the numbers?

New Sets from Old

If A is a set whose elements are from universal set U, then $\{x|x \in U \text{ and } x \notin A\}$ is called the *complement* of A, often denoted A', A^c, or $\overline{A}$. For example, if U is the set of integers and A is the even integers, then A' is the set of odd integers. Note that it is important to be clear on the universal set when finding a complement. The symbol $\emptyset$ stands for the *empty set*, the set with no elements. It follows that $\emptyset' = U$ and $U' = \emptyset$. Also note that $A'' = A$, for every set A. (We leave the proof of this for an exercise.)

> ✓ Suppose A is the set of odd digits. What is the complement of A in the universal set of digits? What is A' in the set of all nonnegative integers? What is A' in the set of all integers?

Given two sets, A and B, we define binary operations $\cup$ (union) and $\cap$ (intersection) as follows:

$$A \cup B = \{x|x \in A \text{ or } x \in B\}, \text{ and}$$
$$A \cap B = \{x|x \in A \text{ and } x \in B\}.$$

For example, if $A = \{2, 3, 5\}$ and $B = \{2, 5, 7\}$ then $A \cup B = \{2, 3, 5, 7\}$ and $A \cap B = \{2, 5\}$. We say A and B are *disjoint* if $A \cap B = \emptyset$. Note that if U is the universal set and A is any set, then $\emptyset \cup A = A$, $U \cup A = U$, $\emptyset \cap A = \emptyset$ and $U \cap A = A$.

> ✓ Suppose A is the set of all even integers and B is the set of all integral multiples of four. Find $A \cap B$. Find $A \cup B$.

We say that A *is contained in* B or A *is a subset of* B, and write $A \subseteq B$, if and only if every element of A is also in B. So, to show that $A \subseteq B$, we need to show that $x \in A$ implies that $x \in B$.

We can now define set equality by saying $A = B$ if and only if $A \subseteq B$ and $B \subseteq A$. Let's restate that: to show that $A = B$, we need to do two things, (1) show that $x \in A$ implies that $x \in B$, and (2) show that $x \in B$ implies $x \in A$.

Associated with a given set A there is another important set, the *power set of* A, denoted $\mathcal{P}(A)$, which is the collection of all subsets of A. Note that the elements of $\mathcal{P}(A)$ are themselves sets. For example, if $A = \{a, b, c\}$, then $\mathcal{P}(A)$ has eight elements: $\mathcal{P}(A) = \{\emptyset, \{a\}, \{b\}, \{c\}, \{a, b\}, \{a, c\}, \{b, c\}, \{a, b, c\}\}$.

In general, if A has n elements, then $P(A)$ will have 2^n elements. We will prove this fact in Chapter 4.

There are some other ways of making new sets from old; these are given in the Exercises.

Properties of Sets

Using the definitions for equality, intersection, and union, it is easy to infer that $A \cap A = A$ and $A \cup A = A$. (Why?) It is also evident from the definition of $\subseteq$ that if $B \subseteq A$, then $A \cup B = A$ and $A \cap B = B$. In particular, if A and C are any two sets, then $(A \cup C) \cap A = A$ and $(A \cup C) \cup A = (A \cup C)$, since $A \subseteq A \cup C$. Also, $(A \cap C) \cup A = A$ and $(A \cap C) \cap A = (A \cap C)$, since $A \cap C \subseteq A$.

Some additional facts about sets follow. We prove one of these facts as an example and ask you to prove the other facts in the Exercises.

- $\cup$ and $\cap$ are commutative. That is,

$$A \cup B = B \cup A \quad \text{and} \quad A \cap B = B \cap A.$$

- $\cup$ and $\cap$ are associative. That is,

$$(A \cup B) \cup C = A \cup (B \cup C) \quad \text{and} \quad (A \cap B) \cap C = A \cap (B \cap C).$$

- $\cup$ distributes over $\cap$ and $\cap$ distributes over $\cup$. That is,

$$A \cup (B \cap C) = (A \cup B) \cap (A \cup C)$$

and

$$A \cap (B \cup C) = (A \cap B) \cup (A \cap C).$$

- For any set A, $A \cup A' = U$ and $A \cap A' = \emptyset$.
- $(A \cup B)' = A' \cap B'$ and $(A \cap B)' = A' \cup B'$. These are known as *DeMorgan's laws.*

Let's prove that $\cap$ distributes over $\cup$. That is, we wish to show that $A \cap (B \cup C) = (A \cap B) \cup (A \cap C)$. Recall that to show two sets equal we must show that each is a subset of the other. We'll first show that $A \cap (B \cup C) \subseteq (A \cap B) \cup (A \cap C)$. We see that,

$$\begin{aligned} x \in A \cap (B \cup C) &\Rightarrow x \in A \text{ and } x \in B \cup C \\ &\Rightarrow x \in A \text{ and } (x \in B \text{ or } x \in C) \\ &\Rightarrow (x \in A \text{ and } x \in B) \text{ or } (x \in A \text{ and } x \in C) \\ &\Rightarrow x \in A \cap B \text{ or } x \in A \cap C \\ &\Rightarrow x \in (A \cap B) \cup (A \cap C), \end{aligned}$$

as desired. To complete the proof, we now need to show that $(A \cap B) \cup (A \cap C) \subseteq A \cap (B \cup C)$. Note that in this case, the reasoning is the reverse of what we just did.

$$
\begin{aligned}
x \in (A \cap B) \cup (A \cap C) &\Rightarrow x \in A \cap B \text{ or } x \in A \cap C \\
&\Rightarrow (x \in A \text{ and } x \in B) \text{ or} \\
&\qquad\quad (x \in A \text{ and } x \in C) \\
&\Rightarrow x \in A \text{ and } (x \in B \text{ or } x \in C) \\
&\Rightarrow x \in A \text{ and } x \in B \cup C \\
&\Rightarrow x \in A \cap (B \cup C)
\end{aligned}
$$

The preceding five facts, along with those given in the preceding paragraphs about the empty set and the universal set, can be used to simplify more complicated expressions. For example, consider the expression $(A \cap B)' \cap A$. We can simplify this in the following steps (with reasons for each step listed):

$$
\begin{aligned}
(A \cap B)' \cap A &= (A' \cup B') \cap A && \text{(DeMorgan's Law)} \\
&= (A' \cap A) \cup (B' \cap A) && (\cap \text{ distributes over } \cup) \\
&= \emptyset \cup (B' \cap A) && (A' \cap A = \emptyset) \\
&= B' \cap A && (\emptyset \cup C = C)
\end{aligned}
$$

You'll agree that $B' \cap A$ is a simpler expression than $(A \cap B)' \cap A$.

A Paradox

It is interesting to note that not all collections of objects are sets. Some collections are not well-defined: the collection of all pretty colors, for example. But there are collections that are not sets for more mathematically fundamental reasons. We show a classic example here.

Recall that what is required of a set is to be able to determine whether or not any given item is an element of the set. That is, if A is a set, and x is any item, it should possible to determine that either $x \in A$ or $x \notin A$. To show that not all collections are sets, we need to show that for some collection of items and some particular item, it is impossible to determine whether or not that item is a member of that collection. We'll use proof by contradiction.

Let's assume that all collections of objects are indeed sets and divide these into two parts: one consisting of those sets that are members of themselves and the other consisting of those sets that are not members of themselves. The latter contains most of the "normal" sets you are familiar with and is apparently quite large. The first collection seems a little strange at first, but there is nothing at first blush prohibiting a set from being an element of itself. Now if all collections are sets, then the second collection (those sets that are not members of themselves) is itself a set. Let's call it B. Now we ask the question "Is B an element of itself?" That is, is $B \in B$?

If the answer is yes, then, by the definition of B (*all* the sets that are not elements of themselves), we must conclude that $B \notin B$, since B must satisfy the requirement for a set to be in B. This is a contradiction, so we conclude that the answer is not yes. On the other hand, if the answer is no, that is, B is not a element of itself, we conclude that B is indeed an element of B, since it now meets the defining requirement for being in this set; again a contradiction. In either case, we arrive at a contradiction. It must be that B is such that we can not always determine if a given set is an element of B or not. But this is exactly what is required of a set; we must be able to tell whether or not an item is in the set. We must conclude that B is not a set. The problem here is, intuitively, that B is just too large. We call collections too large to be sets *proper classes*. We will not deal with proper classes here, only with sets.

Large Collections of Sets

Union and intersection can be extended to more than just two sets. We write the union and intersection of the finite collection of sets $A_1, A_2, \ldots, A_n$ as $\cup_{i=1}^{n} A_i$ and $\cap_{i=1}^{n} A_i$, respectively. The union and intersection of the infinite collection of sets $A_1, A_2, \ldots$ can be written $\cup_{i=1}^{\infty} A_i$ and $\cap_{i=1}^{\infty} A_i$, respectively.

Note that an element is in $\cup_{i=1}^{\infty} A_i$ if it is in any one of the sets A_i and an element is in $\cap_{i=1}^{\infty} A_i$ if it is in *all* of the sets A_i. We can generalize to the union and intersection of any set of sets, regardless how they're indexed. If $\mathcal{A}$ is a set of sets, then $\cup_{A \in \mathcal{A}} A = \{x | x \in A \text{ for some } A \in \mathcal{A}\}$.

We can define the intersection of an arbitrary set of sets in a similar way. For example, let $A_i = [2, 3 + 1/n)$. Then $\cap_{i=1}^{\infty} A_i = [2, 3]$. (Convince yourself this is true.)

We may be interested in the size of a set. This is sometimes referred to as the *cardinality* of a set and denoted $|A|$. As noted previously, if $|A| = n$, then $|\mathcal{P}(A)| = 2^n$. For infinite sets, there are other ways to express cardinality, but we will not cover that topic here.

Exercises

1. Suppose A is the set of distinct letters in the word ELEPHANT. B is the set of distinct letters in the word SYCHOPHANT. C is the set of distinct letters in the word FANTASTIC. D is the set of distinct letters in the word STUDENT. The universe U is the set of 26 capital letters. Find $A \cup B$, $A \cap B$, $A \cap C$, $D \cup A$, $(A \cap C) \cup (B \cap D)$, $A \cap (C \cup D)$, $((B \cup C) \cap (C \cup D))'$, and $(A \cup B \cup C \cup D)'$.
2. Give an explicit representation for the set of all prime numbers less than 50.
3. Give an implicit representation for the set of all prime numbers less than 50.
4. Prove that $\cup$ distributes over $\cap$.
5. Prove that union and intersection are commutative.
6. Prove that $A \cup (A \cap B) = A$.
7. Prove that if $A \subseteq B$, then $B' \subseteq A'$.
8. What can you conclude about A and C if $A \subseteq B$ and $B \subseteq C$? Prove your answer.
9. Prove that if $A \subseteq B$ and $C \subseteq D$, then $A \cup B \subseteq C \cup D$.

10. Prove that for every set A, $\emptyset \subseteq A$.
11. Prove that $A \cap A = A \cup A = A$ for all sets A.
12. Show that $A'' = A$ for any set A.
13. Prove DeMorgan's laws.
14. Prove that $A \cap B \subseteq A \cup B$ for all sets A and B.
15. Prove that the following are equivalent: (1) $A \subseteq B$, (2) $A \cap B = A$, (3) $A \cup B = B$.
16. Recall that sets A and B are *disjoint* if $A \cap B = \emptyset$. If A and B are disjoint, what can you say about the relationship between A, A', B, and B' (in terms of certain sets being subsets of others)? Find as many relations as you can here.
17. Define $A - B = A \backslash B = \{x | x \in A \text{ and } x \notin B\}$. Show that the union of any two sets is the union of disjoint sets. *Hint*: Show that $A \cup B = A \cup (B \backslash A)$. This operation is called *set difference*.
18. Define the universal set $U =$ the set of integers. Let $A = \{x | x \text{ is even}\}$, $B = \{x | x \text{ is odd}\}$, $C = \{x | x < 5\}$. Find $A \cap B$, $A \cup B$, $A \cap C$, $A \cup C$, $C \backslash (A \cup B)$, $C \backslash (A \cup C \backslash B)$, $C \backslash A \cap C \backslash B$, $A \backslash (A \backslash C)$, $U \backslash (A \cap C)$, $U \backslash (A \cup U \backslash C)$, $U \backslash A \cap U \backslash C$, $A \cap B \cap C$.
19. If A and B are sets, prove that $A \subseteq B$ iff $B' \subseteq A'$.
20. Prove that $A \subseteq B$ iff $A \cap B' = \emptyset$
21. If $U = A \cup B$ and $A \cap B = \emptyset$, then $A = U \backslash B$.
22. What can you conclude about the intersection of A and B if $A \subseteq U \backslash B$ and $B \subseteq U \backslash A$? Prove your answer.
23. Suppose for a fixed set A in universe U, $A \cap B = A$ for every set B. What can you conclude about A? Prove your answer.
24. Suppose for a fixed set A in universe U, $A \cup B = B$ for every set B. What can you conclude about A? Prove your answer.

25. The set $A \times B = \{(a, b)|a \in A, b \in B\}$ is called the *Cartesian product* of sets A and B. If $A = \{1, 2\}$ and $B = \{2, 3\}$, find $A \times B$.
26. Prove or disprove: $A \times B = B \times A$. What is $|A \times B|$ in terms of $|A|$ and $|B|$?
27. Is the Cartesian operator $\times$ commutative? If your answer is yes, prove it. If your answer is no, give a counterexample.
28. How would you define $A \times B \times C$? Is $\times$ associative?
29. Describe explicitly the set of all integral solutions to the equation $x^2 + 5x + 6 = 0$.
30. Using H to stand for heads and T to stand for tails, give an explicit representation of the set of all possible outcomes when three coins are tossed. For example, one outcome is HHH.
31. Rewrite the following set using set-builder (implicit) notation: $\{..., -2, 0, 2, 4, ...\}$. Why is the set-builder notation preferred here?
32. Show that $B \backslash (\cup_{A \in \mathcal{A}} A) = \cap_{A \in \mathcal{A}} (B \backslash A)$.
33. Simplify the expressions $(A' \cup B)' \cap (A \cup B)$. Simplify $(A \cup (A' \cup B))'$.
34. List the elements of $\mathcal{P}(\{a, b\})$.
35. Give the power set of $\{x, \{x\}\}$.
36. Which of the following is true for all sets S?
    a. $\emptyset \in \mathcal{P}(S)$
    b. $\emptyset \subseteq \mathcal{P}(S)$
    c. $\emptyset \in S$
    d. $\emptyset \subseteq S$
37. List the elements of $\mathcal{P}(\mathcal{P}(S))$ where $S = \{a, b\}$.
38. What is $|\mathcal{P}(\mathcal{P}(S))|$ when $|S| = n$? (See Exercise 37.)
39. $|A \cup B| = |A| + |B|$ under what conditions?
40. Suppose $A_1 = \{2, 8, 16, 32\}$, $A_2 = \{3, 9, 27\}$, $A_3 = \{2, 3, 12, 19, 27\}$, $A_4 = \{4, 16\}$. Find $\cap_{i=1}^{4} A_i$. Find $\cup_{i=1}^{4} A_i$.

41. Suppose $S_i = \{x | x \in \mathbb{R} \text{ and } 0 \leq x \leq 1/n\}$. Find $\cap_{i=1}^{\infty} S_i$. Find $\cup_{i=1}^{\infty} S_i$.

* 42. Extend DeMorgan's laws to a theorem about $\cup_{i=1}^{n} A_i'$, and for $\cap_{i=1}^{n} A_i'$. Can you do the same for $\cup_{i=1}^{\infty} A_i'$ and $\cap_{i=1}^{\infty} A_i'$?

43. Use set notation to describe the set of all ordered pairs of real numbers such that the second element of the pair is the square of the first element.

44. Does set union have the property of cancellation, i.e. is it true that if $A \cup B = A \cup C$, then $B = C$? Explain your answer.

45. Does set intersection have the property of cancellation?

46. Using theorems of set theory, simplify $(A \cup B) \cap (A' \cup B)$. By *simplify* we mean to write the expression using fewer symbols if possible.

47. Simplify: $A \cap (B \cup (A \cap (B' \cup (A \cap (B \cup (A' \cap B'))))))$.

48. The total number of elements in three sets, A, B, and C is 200. 70 are in A, 120 in B, 90 in C, 50 in $A \cap B$, 30 in $A \cap C$, 40 in $B \cap C$, and 20 in $A \cap B \cap C$. Find how many are in $A \cup B$, $A \cup B \cup C$, $A' \cap B \cap C$, $A \cap B' \cap C'$, and in $A' \cap B' \cap C'$.

49. Prove or disprove: If $A \cup B = A \cup C$, then $B = C$.

50. Prove or disprove: If $A \cap B = A \cup C$, then $B = C$.

51. Prove or disprove: If $A \times B = A \times C$, then $B = C$.

52. Using the definition of Exercise 17, prove or disprove: If $A - B = A - C$, then $B = C$.

53. Prove $(A - B) \cup (B - A) = (A \cup B) - (A \cap B)$.

54. Prove $(A - B) \cap (A - C) = A - (B \cup C)$.

55. Prove $(A \times C) \cap (B \times D) = (A \cap B) \times (C \cap D)$.

56. Under what conditions does $A \cap B = A$? Justify your answer.

57. Under what conditions does $A - B = B - A$? Justify your answer.

58. Prove or disprove: $(A - B) - C = A - (B - C)$.

Programming Problems

1. Devise some method for representing a (finite) set of integers. Write a program that inputs two sets (call them A and B) and then prints A, B, $A \cup B$, $A \cap B$, and $A - B$.
* 2. Write a program to list the elements in the power set of a given (finite) set.
3. Write a program that outputs the number of subsets of each possible size. For example, a set of two elements has one subset of two elements, two subsets of one element each, and one subset with no elements.
4. Write a program that inputs two sets and determines whether one is contained in the other.
5. Write a program that inputs a set and then randomly chooses an element from that set. Write a program that inputs a set and then displays the elements of the set in a random order.
6. Write a program that inputs two sets and then displays their cartesian product.
7. Write a program that defines abstract data type (ADT) *Set*. The operations should include union, intersection, set difference, and complement, as well as some way to input and display sets. Programmers should be able to use your ADT to declare variables of type *Set*.

Chapter 2
Functions and Relations

The idea of associating an element from one set with an element (or elements) from another is a fundamental one in mathematics. There are many different kinds of associations and a wide variety of notations used for expressing them. We will examine two kinds of associations: functions and relations, both of which have significant applications in computer science. We begin with functions, since you are more familiar with them, before considering the more general idea of relations.

Recall that a *function* from A to B is a mapping from one set (the *domain* of the function) to another (the *range* of the function) where each element of the domain is mapped to one element of the range. For example, we could map each element in the set of integers to twice its value. In this case, if we call the function f, one way of expressing the mapping is to write $f(x) = 2x$. It is important to specify the domain and range of a function. For example, if we wanted to restrict the preceding example to the integers, we could write $f : \mathbb{Z} \to \mathbb{Z}$ such that $f(x) = 2x$. Here, $\mathbb{Z}$ stands for the set of integers. On the other hand, we could have a function $g : \mathbb{R} \to \mathbb{R}$ defined by $g(x) = 2x$. ($\mathbb{R}$ is the set of real numbers.) The functions g and f are different but they agree on the set of integers.

The critical idea is that a function maps each element in its domain to a *unique* element in its range. What is not allowed is for an element in the domain to be mapped to more than one element in the range. An example of a mapping that is not a function is one that maps each non-negative real number a to the real numbers x such that $x^2 = a$. This mapping is not a function since, for example, 4 would get mapped to 2 and to -2. (There are many examples where this mapping fails to be a function. If it fails for just one value in the domain, we cannot call the mapping a function.)

We illustrate a mapping that is a function and one that is not in the following diagrams:

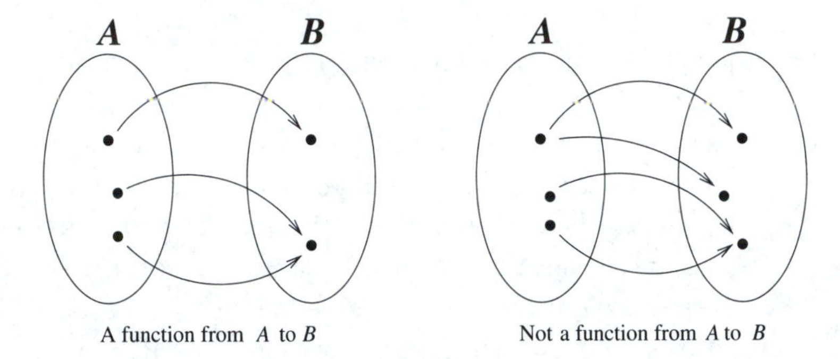

A function from A to B Not a function from A to B

> ✓ Give three representations for the idea of associating any given real number with that number squared plus one. Is this association a function?

There are so many functions that are important to us in computer science that we can't possibly list them all. We'll give a few examples of the most important ones used not only for carrying out computations, but also for analyzing the performance of programs.

Exponential and Log Functions

An important class of functions for us are the exponential functions: $f(x) = b^x$. We call b the *base* of the exponential. Two bases are of particular importance for us: e (Euler's constant $\approx 2.718128\ldots$) and 2. When you graph e^x or 2^x (or any exponential with base larger than 1), you should notice that the function grows very rapidly after a short while. The graphs of all exponential functions, with base greater than 1, have similar shapes. The following graph plots the functions 4^x, e^x and 2^x on one axis. (The function 4^x is plotted with a line of crosses and 2^x is plotted with a dashed line.) Notice that the scales on the two axes are not the same. Also notice how fast these functions grow as the value of x increases, even though we've only shown the values up to $x = 3$.

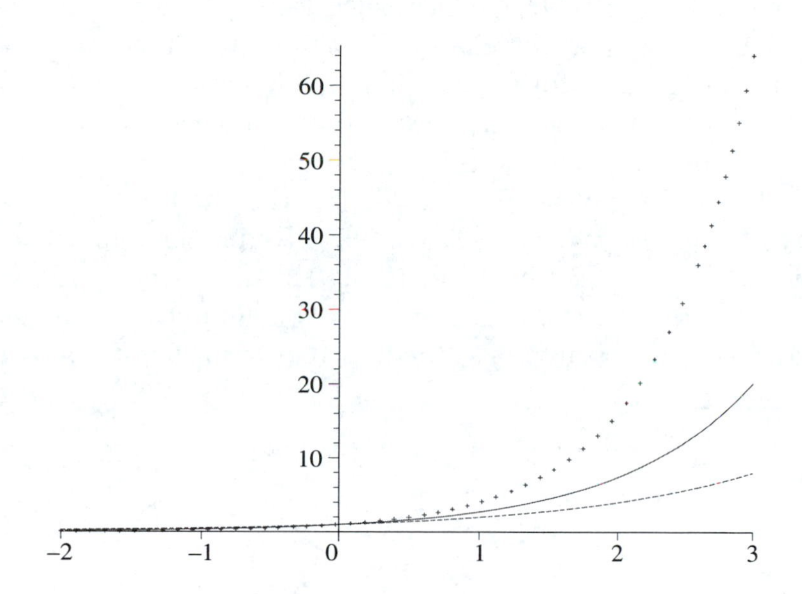

> ✓ Use a symbolic computation software package or a graphing calculator to compare the graphs of the identity function, the function that takes a real number and squares it, and an exponential function that takes any real number and raises 2 to that power.

Recall that a *polynomial over the reals*, p, in the variable x can be thought of as a function of the form $p(x) = a_0 + a_1x + a_2x^2 + \cdots + a_nx^n$, where the coefficients $a_0, \ldots, a_n$ are real numbers. The *degree* of $p(x)$ is the largest n such that $a_n \neq 0$. (If $p(x) = 0$ then we say $p(x)$ has degree -1.) As mentioned, exponential functions grow very rapidly. Indeed, if $p(x)$ is any polynomial and $b > 1$, then there is a number N such that if $x > N$, then $b^x > p(x)$. That is, *eventually* b^x is greater than $p(x)$. This is true regardless of the degree of

$p(x)$ or its coefficients, although for polynomials of particularly large degree, the first time b^x exceeds $p(x)$ (that is, the value of the smallest N mentioned previously) might be a rather large number. For example, $e^x > 1000x^{10} + 3x^3$ for all $x > 45$. (You can check that 45 is the smallest such value where this is true.)

Another important class of functions is the collection of logarithm (or log) functions. The log functions "undo" the mappings of the exponential functions. We need to make this idea of "undoing," or inverse, precise.

The *inverse* of a function f is another function, denoted by f^{-1}, such that $f^{-1}(f(x)) = x$ for all x in the domain of f. Since f^{-1} must itself be a function, f has an inverse only if f is *one-to-one*; that is, only if $f(x) = f(y)$ implies $x = y$. The following are two functions; the first one is not one-to-one and but the second one is.

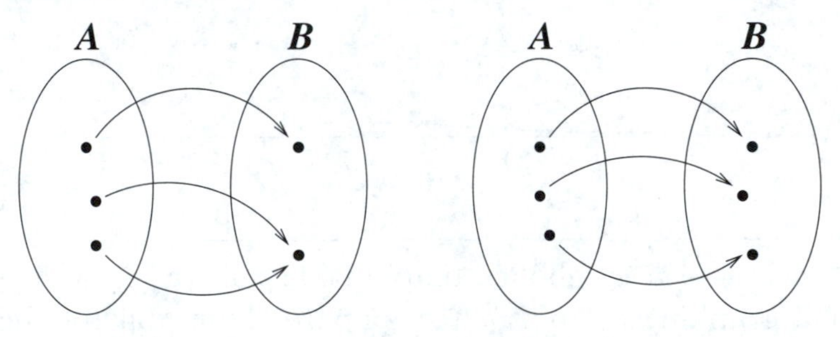

✓	Find the inverse of $f(x) = 3x + 5$.

The inverses of the exponential functions are the logarithm functions. The logarithm base b is written $\log_b(x)$ and so is defined by $\log_b(x) = y$ if and only if $b^y = x$. Thus $\log_2(8) = 3$, since $2^3 = 8$. Note that since the range of an exponential function (with positive base) is the set of positive reals and the domain is the set of all reals, the domain of the corresponding log function is the set of positive reals and the range is the set of all reals. The logarithm base e is called the *natural logarithm* and usually written ln and the logarithm base 2 will be written simply log with no subscript. The following are the graphs of

$\ln x$ and $\log x$. (The graph of $\log x$ is in dashes.)

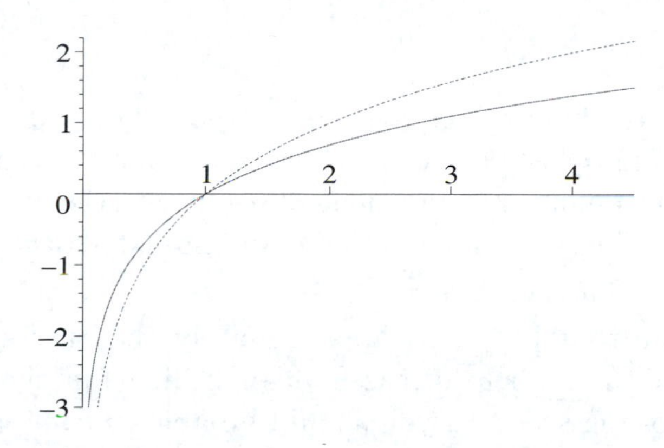

Notice how slowly the log functions grow, in sharp contrast to the exponential functions. Indeed, $\lim_{x\to\infty} \log_b(x) = \infty$, but $\lim_{x\to\infty}$ (slope of $\log_b(x)) = 0$. In other words, the value of $\log_b(x)$ gets as large as you wish, but at a progressively slower rate. Contrast this with the exponential functions, which go to ∞ at a progressively faster rate.

> ✓ Using computer software, compare the graphs of the linear functions $ax + b$ for various a's and b's with the log functions $a \log_b(x)$ for various a's and b's.

Since logs and exponentials are inverses of each other, we have that

$$b^{\log_b(x)} = x, \text{ for } x > 0 \qquad \text{and} \qquad \log_b(b^x) = x, \text{ for all } x.$$

Recall the following properties of exponentials:

$$b^{x+y} = b^x b^y, \qquad b^{x-y} = b^x / b^y, \text{and} \qquad (b^x)^y = b^{xy},$$

which give rise to the corresponding properties of logarithms:

$$\log_b(xy) = \log_b(x) + \log_b(y),$$
$$\log_b(x/y) = \log_b(x) - \log_b(y), \quad \text{and}$$
$$\log_b(x^y) = y \log_b(x).$$

Changing bases of logarithms is simply a matter of dividing by a constant:

$$\log_b(x) = \frac{\log_a(x)}{\log_a(b)}.$$

That is, to change from $\log_a$ to $\log_b$, simply divide by $\log_a(b)$. So, $\log_3(x) = \log_2(x)/\log_2 3$. This last property is easily shown from the definition and elementary properties of log: Suppose $y = \log_b(x)$. Then $b^y = x$ and so $\log_a(x) = \log_a(b^y) = y\log_a(b)$. Therefore, $y = \frac{\log_a(x)}{\log_a(b)}$.

Note that $\log_{10} n$ is approximately the number of digits in n. With the help of the floor or ceiling function (given in the following section), you should be able to come up with an exact formula. (See the Exercises.)

Floor and Ceiling Functions

Two other examples of useful functions are the floor and ceiling functions, written $\lfloor x \rfloor$ and $\lceil x \rceil$, respectively, which are defined for all real x as follows:

$$\lfloor x \rfloor = \text{the largest integer less than or equal to } x$$
$$\lceil x \rceil = \text{the smallest integer greater than or equal to } x$$

Thus, $\lfloor 2.7 \rfloor = 2$ and $\lceil 2.7 \rceil = 3$. Note that $\lfloor x \rfloor = \lceil x \rceil$ if and only if x is an integer, in which case the value of both of these functions is x. Note the following for integer n and real x:

$$\lfloor x \rfloor = n \text{ if and only if } n \le x < n+1,$$
$$\lfloor x \rfloor = n \text{ if and only if } x-1 < n \le x,$$
$$\lceil x \rceil = n \text{ if and only if } n-1 < x \le n,$$
$$\lceil x \rceil = n \text{ if and only if } x \le n < x+1.$$

Sometimes we call the floor of x the integer portion of x and $x - \lfloor x \rfloor$ the fractional portion of x. Note that the truncation function available in most programming languages (for instance, the `trunc` function in Pascal) is the floor function.

Neither the floor nor the ceiling function is additive or multiplicative. That is, it is not always the case that $\lfloor x+y \rfloor = \lfloor x \rfloor + \lfloor y \rfloor$ and similarly it is not always true that $\lceil x+y \rceil = \lceil x \rceil + \lceil y \rceil$; this is also the case for multiplication. (This is left as an exercise.)

Note that integer division can be written $\lfloor n/m \rfloor$, where n and m are integers. Thus $n \bmod m = n - \lfloor n/m \rfloor m$. Most programming languages use the floor function applied to the quotient for integer division.

> ✓ Compare the integer division of your favorite programming language with the formula given previously using the floor function to see whether or not your language uses the floor function for doing integer division. Test both versions for various values of n and m.

Relations

We have seen that a function associates elements of one set (the domain) with elements of another set (the range). But these associations must satisfy a very special requirement: each element of the domain is associated with exactly one element in the range. A more general association is called a relation.

The *Cartesian product* of A and B, denoted $A \times B$, is the set of ordered pairs $\{(a, b) : a \in A, b \in B\}$. Any subset of $A \times B$ is called a *relation* from A to B. For example, if $A = \{1, 2, 3\}$ and $B = \{2, 4\}$, consider $R = \{(1, 2), (1, 4), (2, 2), (2, 4), (3, 4)\}$. It is possible to observe that the relation R just given is the "less than or equal to" relation.

> ✓ $A = \{a, b\}$ and $B = \{c, d, e\}$. Find five different relations from A to B.

Some relations have special properties. For example, functions are relations. It is sometimes worthwhile to think of a function as a set of ordered pairs. Indeed, if D is the domain and R is the range of a function f, then we can think of f as a set of ordered pairs (d, r), where $d \in D$ and $r \in R$ with the property that if $(d, r_1) \in f$ and $(d, r_2) \in f$, then $r_1 = r_2$. This is simply restating, in the language of relations, the critical defining fact about functions: A value in the domain gets mapped to a unique element of the range. Thinking of a function in this way, the function on the integers $f(x) = x^2$ would be

$$f = \{\ldots, (-2, 4), (-1, 1), (0, 0), (1, 1), (2, 4), (3, 9), \ldots\}.$$

This way of expressing a function could obviously be cumbersome or even impossible, but sometimes it is convenient.

If we look at a function as a set of ordered pairs, the inverse of a function is just the pairs with the coordinates reversed. That is, if $f = \{(a, b) | a \in D, b \in R\}$, then $f^{-1} = \{(b, a) | a \in D, b \in R\}$, provided f is one-to-one. That is, $(a_1, b) \in f$ and $(a_2, b) \in f$ implies that $a_1 = a_2$. Note that when viewing f as a set of ordered pairs it is easy to see why f needs to be one-to-one in order for f^{-1} to be a function. For if f were not one-to-one, $(a_1, b) \in f$ and $(a_2, b) \in f$, where $a_1 \neq a_2$, then $(b, a_1) \in f^{-1}$ and $(b, a_2) \in f^{-1}$ and so f^{-1} would not be a function, since b would get mapped to two distinct values.

> ✓ For the sets A and B in the last check box, find all functions from A to B.

While we can have relations between elements of two different sets, frequently we are interested in a relation between elements of the same set. For any set A, a *relation* on A is a subset $\mathcal{R}$ of $A \times A$. ($A \times A$ is the set of ordered pairs (a, b) where $a, b \in A$.) For example, if $A = \{1, 2, 3\}$ and $\mathcal{R} = \{(1, 2), (2, 3), (1, 3)\}$, then $\mathcal{R}$ is the usual "less than" relation. This method of expressing a relation as a set of ordered pairs is correct, but frequently a more common method is that of using a symbol between two elements of A. For example, we usually write $1 < 2$ and $2 < 3$ to express the relation $\mathcal{R}$ we just gave. Sometimes it is convenient to think of a relation as ordered pairs, sometimes not. We will choose the more appropriate notation, depending on the circumstances.

Relations may enjoy certain properties. A relation $\mathcal{R}$ on the set A is *reflexive* if $(a, a) \in \mathcal{R}$ for each $a \in A$. Evidently, the relation $<$ on the reals is not reflexive, while $\leq$ is.

$\mathcal{R}$ is *symmetric* if $(a, b) \in \mathcal{R}$ implies that $(b, a) \in \mathcal{R}$. Neither $<$ nor $\leq$ on the reals is symmetric. For example, $3 < 4$ but $4 \not< 3$. But the relation $\sim$ on the integers given by $a \sim b$ if $a - b$ is even is symmetric, since $a - b$ is even implies that $b - a$ is also (being negations of each other). Thus $a \sim b$ implies $b \sim a$. The relation "is similar to" on the set of triangles is symmetric.

$\mathcal{R}$ is *antisymmetric* if $(a, b) \in \mathcal{R}$ and $(b, a) \in \mathcal{R}$ implies $a = b$. Said another way, $\mathcal{R}$ is *antisymmetric* if whenever $(a, b) \in \mathcal{R}$ and $a \neq b$, then $(b, a) \notin \mathcal{R}$. Clearly, $\leq$ is antisymmetric but the relation $\sim$ just given is not. To see that $\sim$ is not antisymmetric, we see that $3 \sim 7$ (since $3 - 7 = -4$) but also $7 \sim 3$. There are many other such examples where the relation fails to be antisymmetric. Keep in mind that you only need one counterexample to show that the relation is not antisymmetric.

Finally, $\mathcal{R}$ is *transitive* if whenever $(a, b) \in \mathcal{R}$ and $(b, c) \in \mathcal{R}$, then $(a, c) \in \mathcal{R}$. All three relations just given are transitive, as you can check. However, let $x \sim y$ if $|x - y| < 1$. This relation on the reals is not transitive, since $1 \sim 1.8$ and $1.8 \sim 2.5$ but $1 \not\sim 2.5$. The relation is symmetric, however, as you can check.

If a relation is reflexive, symmetric, and transitive, we call it an *equivalence relation*. Frequently, the symbol $\equiv$ is used for an equivalence relation because an equivalence relation acts much like equality. (By the way, the relation "equal" is an equivalence relation on the real numbers.) Another example of an equivalence relation on the integers is $a \equiv_2 b$ if $a - b$ is even. Note that $\equiv_2$ is reflexive ($a - a = 0$), symmetric (if $a - b$

is even, so is $b - a$), and transitive (if $a - b$ and $b - c$ are even, then $a - c = (a - b) + (b - c)$ is also, being the sum of two even integers).

> ✓ If $A = \{a, b, c\}$, find all equivalence relations on A.

Note that $\equiv_2$ partitions the integers into two disjoint sets (called equivalence classes), the even integers and the odd integers. To describe it another way, the two classes are those integers that are equivalent to 0 and those that are equivalent to 1. Similarly, if we defined the relation $\equiv_3$ by $a \equiv_3 b$ if $a - b$ is a multiple of 3, then $\equiv_3$ is an equivalence relation. Furthermore, $\equiv_3$ partitions the integers into three classes, those integers equivalent to 0, those integers equivalent to 1, and those integers equivalent to 2. This partitioning always happens with equivalence relations.

Let's state the preceding explicitly: *If $\equiv$ is an equivalence relation on the set A, then A is partitioned into equivalence classes such that a and b belong to the same class if and only if $a \equiv b$.* This is easy to show. A *partition* of a set is a collection of disjoint subsets whose union is the entire set. The equivalence class for $\equiv$ that contains a (let's call this set E_a) is all the $b \in A$ such that $a \equiv b$. First note that every element a of A is in some class, namely E_a, since $\equiv$ is reflexive. Now all we must show is that if E_a and E_b overlap, they are identical. But if $c \in E_a \cap E_b$, then $a \equiv c$ and $b \equiv c$. But then $c \equiv b$ (since $\equiv$ is symmetric) and so $a \equiv b$ (since $\equiv$ is transitive). That is, $b \in E_a$. It follows that $E_b \subseteq E_a$. (Why? If $c \in E_b$, then $b \equiv c$. But $b \in E_a$ and so $a \equiv b$. Since $\equiv$ is transitive, $a \equiv c$ and so $c \in E_a$.) A similar argument shows that $E_a \subseteq E_b$ and so $E_a = E_b$.

The simplest equivalence relation "equals" on any set has rather trivial equivalence classes. Indeed, $E_a = \{a\}$, for element a. At the other extreme, consider the equivalence relation on any set where every element is related to every other element. How many equivalence classes are there for this relation?

Another special type of relation is one that is reflexive, antisymmetric, and transitive. Such a relation is called a *partial order*. The canonical example of a partial order is the relation "subset" on the power set of a given set. That is, the set A is related to the set B if $A \subseteq B$. In the Exercises, you will be asked to verify this. Note that when considering this relation on all subsets of a given set with more than one element, not every pair of subsets are related. For example, if we consider the subset relation on the power set of $\{1, 2, 3, 4, 5\}$, then $\{1, 2, 3\}$ and $\{2, 4\}$ are not related. However, a partial order where each pair is related is called a *total order*. Note that $\leq$ on the reals (or on any subset of the reals, such as the integers) is a total order. To verify this you must first check that it is a partial order, then check that every pair of elements is related.

When writing a program, you may wish to arrange items in a special order that is different from a usual numeric or alphabetical order. You may wish to sort a collection of records where no order is obvious or where the order is more complex than, say, Social Security numbers or zip codes. You must first establish an order. That is, indicate what it means for one record to be "greater than" another. When establishing this relation, you must guarantee it is indeed at least a partial order, and probably a total order. See the Exercises for some additional examples of orders.

Finally, the examples we've given naturally arise from some well-known concepts. But recall that a relation on a set A is *any* subset of $A \times A$. So if A is finite and large or infinite, you have a large number of possible relations. Thus you may have a particular relation in mind, but no easy way to describe it.

Exercises

1. Graph the functions $y = e^x$, $y = 2^x$, $y = 3^x$, $y = \ln(x)$, $y = \log_2(x)$, and $y = \log_3(x)$.
2. Find the smallest integer N so that $2^x > 2x^3 + x^2 + 1$ for

all $x > N$.

3. Show that if n is an integer, then
   i. $x < n$ if and only if $\lfloor x \rfloor < n$,
   ii. $n < x$ if and only if $n < \lceil x \rceil$,
   iii. $x \leq n$ if and only if $\lceil x \rceil \leq n$, and
   iv. $n \leq x$ if and only if $n \leq \lfloor x \rfloor$.
4. Give instances of x and y that make the following true and instances that make the following false.

$$\lfloor x + y \rfloor = \lfloor x \rfloor + \lfloor y \rfloor, \qquad \lceil x + y \rceil = \lceil x \rceil + \lceil y \rceil$$

   Can you replace the equal signs with inequalities of some sort so the statements will always be true? Repeat the exercise for multiplication.
5. Give a precise formula for the number of digits in an integer n.
6. While most programming languages have some sort of built-in floor function, not all have a ceiling function. Write one.
7. Assuming only the floor function is provided, write a round function.
8. Show that the relation $\sim$ on the reals given by $x \sim y$ if $|x-y| < 1$ is not transitive. Is it reflexive? Is it symmetric? Is it antisymmetric?
9. Define a relation on the reals that is reflexive but not symmetric or transitive; one that is symmetric but not transitive or reflexive; one that is transitive but not reflexive or symmetric.
10. If T is the set of all triangles, let triangle A be related to triangle B if A is similar to B. Is this an equivalence relation? Why or why not?
11. For the equivalence relation *equals* on the real numbers, what is E_a for each real number a?

12. Consider the relation on the reals where every real is related to every other real. Describe the equivalence class E_a for each real number a.

13. For each relation, indicate if it is reflexive, symmetric, antisymmetric, or transitive. Which of the following are equivalence relations? If it is an equivalence relation, list (or describe) the equivalence classes.

    On the integers: $a \sim b$ if $a \geq b$.
    On the integers: $a \sim b$ if $a = b + 1$ or $a = b - 1$.
    On the reals: $a \sim b$ if $|a - b|$ is an integer.
    On the reals: $a \sim b$ if $b = a^n$ some integer n.
    On all rectangles: $a \sim b$ if area of a = area of b.
    On all rectangles: $a \sim b$ if one side of a has same length as one side of b.
    On all people: $a \sim b$ if a and b have the same parents.
    On all people: $a \sim b$ if a and b have at least one parent the same.
    On all college grads: $a \sim b$ if a and b are alums of the same school.
    On all college grads: $a \sim b$ if a and b had the same major (allowing double majors).

14. Show that $\subseteq$ is a partial order on the power set of a given set S.

15. Show that $\leq$ is a total order on the reals.

* 16. Let $\mathbb{Z}$ be the set of integers and let $\mathbb{Z} \times \mathbb{Z} = \{(a, b) : a, b \in \mathbb{Z}\}$ be the set of ordered pairs with coordinates from the integers. Now set $(a, b) \leq (c, d)$ if $a < c$ or $a = c$ and $b \leq d$. Show that $\leq$ is a total order on $\mathbb{Z} \times \mathbb{Z}$.

* 17. Here's another relation on $\mathbb{Z} \times \mathbb{Z}$: set $(a, b) \leq (c, d)$ if $a \leq c$ and $b \leq d$. Show that $\leq$ is a partial order. Is it a total order?

18. For the sets $A = \{x, y\}$ and $B = \{r, s, t\}$,
    a. find all functions from A to B.
    b. find all functions from B to A.

c. find all relations on A.
d. find two relations from A to B that aren't functions.
e. find how many different relations there are from A to B. (Can you generalize?)
f. find a relation on B that is a equivalence relations. Find one that is a partial order. Find one that is a total order.

19. Can an equivalence relation be an order relation? Why or why not?

20. Can an order relation be an equivalence relation? Why or why not?

21. Is every function a relation? Why or why not?

22. Is every relation a function? Why or why not?

23. Give an example of a relation that is reflexive, but not symmetric and not transitive.

24. Give an example of a relation that is symmetric, but not reflexive and not transitive.

25. Give an example of a relation that is transitive, but not reflexive and not symmetric.

26. Suppose you have a collection of computer network nodes and you know which nodes have direct connections to which other nodes. For example, if the nodes are called A, B, C, D, and E, you might know that there are connections between A and B, between A and C, between D and B, between D and E, between A and E, between C and D. Show that the relation on the set of nodes defined by "can communicate with" is an equivalence relation.

* 27. Consider the following definition about functions: Given functions f and g, f is called "big Oh of g," written $f = O(g)$, iff there exist constants n_0 and c, n_0 a natural number and c a real number, such that $f(n) = cg(n)$ for all $n > n_0$. Using this definition, answer the following questions:

a. If $f(x) = x$ and $g(x) = x^2$, is it the case that $f = O(g)$? Does $g = O(f)$?
b. If $f(x) = \log(x)$ and $g(x) = x$, does $f = O(g)$? Does $g = O(f)$?
c. If $f(x) = 3x^2+5x+12$ and $g(x) = x^2$, does $f = O(g)$? Does $g = O(f)$?
d. If $f(x) = \sqrt{x}$ and $g(x) = \log(x)$, does $f = O(g)$? Does $g = O(f)$?

28. Suppose f and g are functions. f has domain A and range B. g has domain B and range C. Define the *composition* of f and g, written $g \circ f$, to mean the function that first applies f to a value in A (which results in a value in B) and then applies g to that value. Suppose f is a function from integers to integers and that $f(n) = 2n$. Suppose g is a function from integers to rationals and that $g(n) = 1/n$. Find $(g \circ f)(7)$. Find $(g \circ f)(x)$ for any integer x. Is $f \circ g$ defined? Explain.

29. Suppose f and g are both functions from reals to reals. $f(x) = 5x + 3$ and $g(x) = 6 - x$. Find $g \circ f$. Find $f \circ g$.

30. Find the inverse of $g \circ f$ when it exists.

31. Show that when defined, $(h \circ g) \circ f = h \circ (g \circ f)$, assuming that f, g, and h are functions. *Note*: This shows that $\circ$ is associative.

32. Define relation R by xRy iff $x^2 = y^2$. Show that R is an equivalence relation. What are the equivalence classes defined by R?

33. A function $f : A \rightarrow B$ is *onto* if for each $b \in B$ there is an $a \in A$ such that $f(a) = b$. If a function is one-to-one, it is sometimes called *injective*. If a function is onto it is sometimes called *surjective*. For each of the five following functions, determine if it is injective and if it is surjective. Find the inverse if it exists.
$f(x) = 5x$
$f(x) = \frac{-1}{|x|+1}$

$f(x) = x^2 + 1$
$f(x) = 2 - 3x$
$f(x) = x^{\frac{1}{2}}$

34. Suppose A has m elements and B has n elements, where $m \leq n$. How many one-to-one functions are there from A to B?

35. Suppose $f : A \to B$ and $g : B \to C$ are both injective. Prove that $g \circ f$ is injective.

36. Suppose $f : A \to B$ and $g : B \to C$ are both surjective. Prove that $g \circ f$ is surjective.

Programming Problems

1. Write a program that inputs two polynomials and prints their sum, difference, and product. You will need first to decide how to store a polynomial.

2. Write a program to compute the derivative of a polynomial. You will need first to decide how to store a polynomial.

* 3. Write a program that prints out all the functions on a given set S. Here the input will be the elements in a set.

Chapter 3
Boolean Algebra

Every area of mathematics and computer science depends in some way on logic and the language of sets. We have already developed some notation for expressing ideas about sets. In this chapter we will examine some concepts and properties of sets, and we will make some observations about analogous concepts and properties of propositional logic.

After studying properties of sets and properties of propositions, we will generalize from the two examples to provide a set of axioms encompassing those properties common to both. The resulting structure is called a *Boolean algebra.*

Interestingly enough, it turns out that switching circuits, together with operations for connecting them, form a Boolean algebra. This means that when designing a computer circuit that carries out some operation such as addition, we can reason about the circuit (and perhaps simplify it) by applying the axioms and theorems of Boolean algebra, the abstraction that encompasses sets, propositions, switching circuits and other systems as well.

Propositional Logic

You should review propositional logic from Chapter 0. To recap: Propositions have a value of t (true) or f (false). The truth symbol T is the proposition that's always true and F is the proposition that is always false. Propositional variables, p, q, $r, \ldots$, and truth symbols may be combined by the connectives $\neg$ (not), $\wedge$ (and), and $\vee$ (or). (There is also the connective $\Rightarrow$ (implies) but that will not play a big role here.)

The function connectives are given in the following truth table:

p	q	$\neg p$	$p \wedge q$	$p \vee q$
t	t	f	t	t
t	f	f	f	t
f	t	t	f	t
f	f	t	f	f

For our purposes, the following are the important facts about propositional logic. Instead of the equivalence sign ($\equiv$) we'll use the equal sign ($=$) here.

- $\wedge$ and $\vee$ are commutative. That is, $p \wedge q = q \wedge p$ and $p \vee q = q \vee p$.
- $\wedge$ and $\vee$ are associative. That is, $p \wedge (q \wedge r) = (p \wedge q) \wedge r$ and $p \vee (q \vee r) = (p \vee q) \vee r$.
- $\vee$ distributes over $\wedge$. That is, $p \vee (q \wedge r) = (p \vee q) \wedge (p \vee r)$.
- $\wedge$ distributes over $\vee$. That is, $p \wedge (q \vee r) = (p \wedge q) \vee (p \wedge r)$.
- T acts as an identity for $\wedge$ and that F acts as an identity for $\vee$. That is, $T \wedge p = p$ and $F \vee p = p$.
- $p \wedge (\neg p) = F$ and $p \vee (\neg p) = T$.

Sets

You might want to review quickly the preliminary chapter on sets, Chapter 1. We will be using the terms and notation from that chapter here. We will see a connection between $\vee$ and $\wedge$ of propositional logic and $\cup$ and $\cap$ of set theory. We will consider a given set S as a universal set and all the subsets of S (the power set of S, $\mathcal{P}(\mathcal{S})$) together with $\cup$, $\cap$, and complementation.

Given two sets, A and B, recall that we defined binary operations $\cup$ and $\cap$ as follows:

$$A \cup B = \{x | x \in A \text{ or } x \in B\}, \text{ and}$$
$$A \cap B = \{x | x \in A \text{ and } x \in B\}.$$

"$x \in A$" is a proposition as is "$x \in B$." Thus the union of sets A and B can be characterized by the proposition "$x \in A$ or $x \in B$."

We noted the following properties of sets in the first chapter. Notice that these are similar to the properties we listed for propositional logic.

- $\cup$ and $\cap$ are commutative and associative.
- $\cup$ distributes over $\cap$ and $\cap$ distributes over $\cup$.
- $\emptyset$ serves as an identity for $\cup$ and the universal set U serves as an identity for $\cap$.
- For any set A, $A \cup A' = U$ and $A \cap A' = \emptyset$.

Boolean Algebras

In the late nineteenth century, George Boole noticed the commonalities shared by collections of sets and collections of propositions. He defined an abstraction, an algebra, with those common properties. His abstract system has since been applied in many areas of mathematics and computer science. The abstraction is now called a *Boolean algebra.* We will give a definition of this abstraction and then see some examples in addition to sets and propositions where computer scientists use this abstraction. These common properties are collected into a set of axioms.

Let B represent a set with two binary operations $+$ and $*$. Then B is a Boolean algebra if the following axioms hold:

Axiom 1. Both $+$ and $*$ are commutative. That is, for all a and b in B, $a + b = b + a$ and $a * b = b * a$.

Axiom 2. There exist two special elements of B, 0 and 1, such that for any a in B, $a+0 = a$ and $a*1 = a$. That is, 0 acts as an identity for $+$ and 1 acts as an identity for $*$.

Axiom 3. For each element a in B there is special element a' satisfying such that $a + a' = 1$ and $a * a' = 0$. This element is called the *dual* of a.

Axiom 4. Each operation distributes over the other; that is, for any a, b and c in B, $a * (b + c) = a * b + a * c$ and $a + b * c = (a + b) * (a + c)$.

Here, the symbols 0 and 1, + and $*$ are thought of as just that, symbols, and must not be confused with their use in well-known algebraic systems such as integers and reals. We know that in mathematics we often use a single symbol to represent more than one concept. Usually, it is not difficult to figure out what the interpretation of a symbol ought to be by viewing it in a given context.

For example, if we see $5+8$, we interpret the + as that of integers. If we see $2.75+e$, we interpret the + as that of reals. And if we see $2 * i + 7$, we interpret the + as that of complex numbers. Computer scientists refer to this use of a symbol for multiple purposes as "overloading."

It doesn't take long for us to see that the set of integers, together with addition and multiplication and the usual 0 and 1, will not satisfy the axioms for a Boolean algebra. For example, in checking Axiom 4, we note that although multiplication distributes over addition, addition does not distribute over multiplication; for example, $2 + 3 * 5$ is not the same as $(2+3) * (2+5)$.

To see that such a system can really exist—that is, to show consistency of this set of axioms—we need an example of a set with two operations satisfying the four axioms. Consider the power set $\mathcal{P}(\mathcal{S})$, of a given set S, together with operations union and intersection. Let's have a look at the axioms one by one.

Axiom 1: Both union and intersection are commutative.

Axiom 2: Let the empty set be 0 and the entire set S be 1. For any set A, $A \cup \emptyset = A$ and $A \cap S = A$.

Axiom 3: Given a set A, let A' be the complement of A in S. Then $A \cup A' = S$ and $A \cap A' = \emptyset$.

Axiom 4: Union distributes over intersection and intersection distributes over union.

You have proven these four properties in past exercises and so we have now established that there is at least one Boolean algebra, namely, the power set of a given set under $\cup$ and $\cap$.

It is also straightforward (and left as an exercise) to see that propositional logic is a Boolean algebra. To see this, you must decide what operation should be $+$ and what should be $*$, what propositions should be 0 and 1, what the dual of a proposition should be and that the four axioms are satisfied.

> ✓ How many elements will be in a Boolean algebra that consists of the subsets of a given finite set?

Some Boolean Algebra Theorems

One of the advantages of defining an abstract mathematical system is that we may discover and prove a variety of theorems for the abstract system and then apply those theorems to all of the models of that system. In this section we will examine several theorems and later we will apply those theorems. We will number the theorems so that they will be easy to refer to when we use them. In each of these theorems, B is a Boolean algebra.

Theorem 1. *For all a in B, $a + a = a$.*

Proof:

$$\begin{aligned} a &= a + 0, && \text{by Axiom 2} \\ &= a + a * a', && \text{by Axiom 3} \\ &= (a + a) * (a + a'), && \text{by Axiom 4} \\ &= (a + a) * 1, && \text{by Axiom 3} \\ &= a + a, && \text{by Axiom 2} \qquad // \end{aligned}$$

The property demonstrated by Theorem 1 is called the *idempotent property*; here we say "every element is idempotent under $+$." Note that every step in the proof can be justified by one of the axioms of Boolean algebra. The technique used is a common one. We start with one side of an equation we want to establish and then, using known facts (axioms, definitions, or other theorems), we proceed to get the other side.

Theorem 2. *For all a in B, $a * a = a$.*

Proof:

$$\begin{aligned} a &= a * 1, && \text{by Axiom 2} \\ &= a * (a + a'), && \text{by Axiom 3} \\ &= a * a + a * a', && \text{by Axiom 4} \\ &= a * a + 0, && \text{by Axiom 3} \\ &= a * a, && \text{by Axiom 2} \quad // \end{aligned}$$

Thus every element is also idempotent under $*$. Note that Theorem 2 looks like Theorem 1 with all the $+$ signs replaced by $*$ signs and all $*$'s replaced with $+$'s and that 0's and 1's are interchanged. We call Theorem 2 the *dual* of Theorem 1. We can form the dual of many theorems in Boolean algebra by the same method. If a theorem about Boolean algebras is true so is its dual. (An unfortunate conflict of terminology occurs with *dual.* We now have the dual of an element and the dual of a theorem. The context will always make clear which one we're using, but you should be aware of this as you start your study of Boolean algebras.)

The following are additional theorems true for all Boolean algebras. All of these theorems come in pairs; the dual of each is also a theorem and is given alongside. We leave the proofs of these theorems and their duals to the Exercises. In each theorem, B is a Boolean algebra.

Theorem 3. For all a in B, $a + 1 = 1$. (Dual: $a * 0 = 0$.)

Theorem 4. For all a and b in B, $a + (a * b) = a$. (Dual: $a * (a + b) = a$).)

Theorem 5. For all a, b and c in B, $(a + b) + c = a + (b + c)$. (Dual: $(a * b) * c = a * (b * c)$.) That is, $+$ and $*$ are *associative.*

Theorem 6. For all a in B, $(a')' = a$. (Dual: same statement.)

Theorem 7. $0' = 1$. (Dual: $1' = 0$.)

Theorem 8. For all a and b in B, $(a*b)' = a' + b'$. (Dual: $(a + b)' = a' * b'$.) These are known as *DeMorgan's laws.*

Switching Circuits

We turn now to a Boolean algebra application of special interest to computer science. At the lowest level of digital computers are circuits whose pathways (such as wires or paths on circuit boards) each have one of two values. One way to think of these values is to think of the pathway as being turned on or off. Another way is to think of these values as having low voltage or high voltage. We call the value of these pathways *bits.* We will indicate the value of these pathways with a 0 (for "off" or "low voltage") and a 1 (for "on" or "high voltage"). Pathways may pass through devices called gates. Gates have one, two or more input pathways and one output, whose value is determined by the values of its inputs. Gates come in different types. We'll start with the three basic types of gates: the *and* gate, the *or* gate and the *not* gate.

An *and* gate takes two inputs and produces one output. The following table describes how the *and* gate functions:

input1	input2	output
0	0	0
0	1	0
1	0	0
1	1	1

Note the similarity to the truth table for joining two logical propositions with an *and* (the logical operator $\wedge$). We draw an *and* gate as follows:

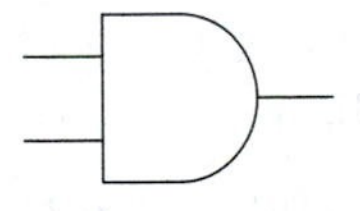

Similarly, we define an *or* gate with the following table:

input1	input2	output
0	0	0
0	1	1
1	0	1
1	1	1

Here we see a strong similarity to an *or* (the logical operator $\vee$) for logical propositions. The symbol for an *or* gate is:

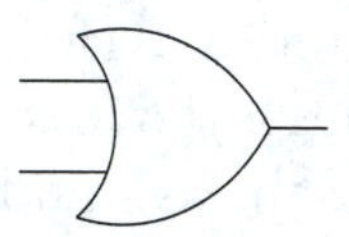

A *not* gate can be defined by the following table:

input	output
0	1
1	0

It is not surprising to see a similarity between the *not* gate and the logical not operator, $\neg$. The symbol for a *not* gate is

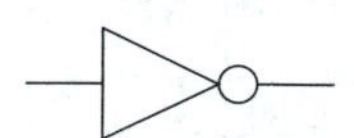

By now you may have guessed that if we associate the value 1 of switching circuits with T of propositional logic and the value 0 with F, then the *and* gate and the *and* logical operator correspond as do the *or* gate and the *or* logical operator and the *not* gate and the *not* logical operator. Indeed switching circuits behave exactly like propositional logic, a fact we'll show soon by showing that switching circuits with these natural operations form a Boolean algebra. In fact, sometimes switching circuits are called logic circuits.

Suppose that A and B are two bits. The algebraic (propositional logic) notation for the output of an *or* gate with inputs A and B is $A + B$. The algebraic notation for the output of an *and* gate with inputs A and B is simply juxtaposition: AB. The algebraic notation for the output of a *not* gate with input

A is A'.

More complicated switching circuits can be expressed with longer expressions. For instance, the switching circuit expression $A + (BC')'$ can be drawn as

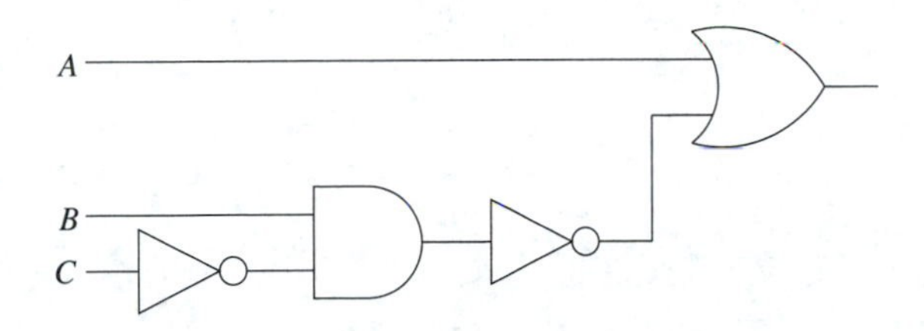

Conversely, when given a more complex switching circuit, we can easily find its expression. For example, the following circuit,

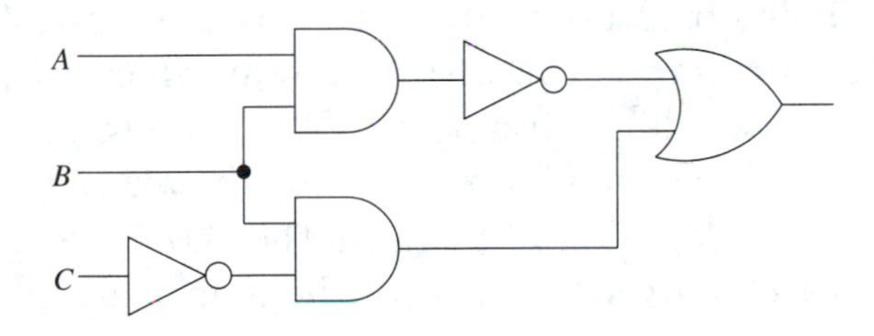

has the expression $(AB)' + BC'$.

Often when designing a circuit, we may find multiple ways to put together gates to satisfy the requirements of the given circuit. For a simple example, the circuit $A + A$ gives no more than the simpler circuit A. When presented with a complex circuit, we'd like to find a simpler one, if possible. This frequently makes the circuit more efficient or less expensive to build. If we know that switching circuits form a Boolean algebra, we can use our Boolean algebra theorems to simplify the expression for the circuit.

For example, consider the following circuit diagram:

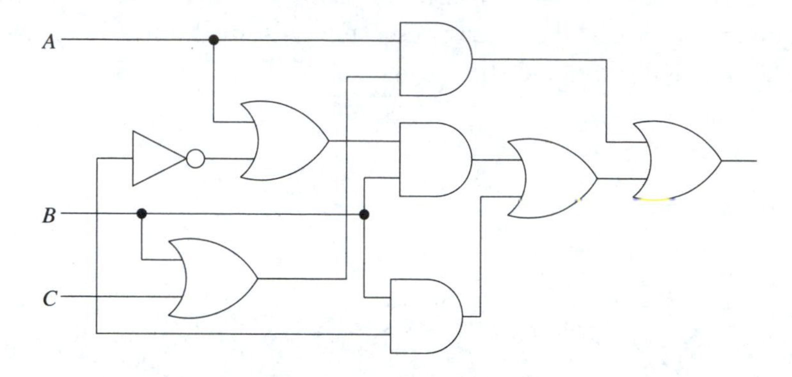

This circuit has eight gates and three inputs. We wonder if it might be possible to simplify the circuit to get an equivalent circuit with fewer gates. We have already noted several similarities between circuits with switching gates and logical propositions. This observation gives us reason to investigate just how much in common these two systems share. Indeed, we can show that switching circuits are Boolean algebras.

For example, if we examine the circuit drawn in the preceding figure, we can represent it by using the expression $A(B+C)+B(A+C')+CB$ given in terms of the names of the input bits, using capital letters such as A to represent those bits.

To show we have a Boolean algebra here, we must designate what the two operations are and decide what the complement of A should be. As our notation suggests, we will let the *or* gate (+) be addition in the Boolean algebra, the *and* gate (juxtaposition) be the multiplication, and the *not* gate ($'$) to be complementation. In order to check that we do indeed have a Boolean algebra we must also provide special gates that act as the "0" and "1" and show that the axioms hold.

It is clear that given any two bits, A and B, connected through an *or* gate with A first, then B, gives, by definition, the same value as putting B first, then A; and similarly for bits connected through an *and* gate. Thus Axiom 1 for Boolean algebras holds.

Since the 0 must act as an additive identity, we need a bit that when we *or* it with any other bit A will yield the value of A. Here we consider a permanently open (or off) bit using 0 as the symbol for it. It is easy to see that this value fills the bill.

Analogously, for the 1, we identify a bit that is always closed (or on). Consider the circuit in which we *and* any bit A with 1; the result is A. Hence, the 1 acts as an identity for the and gate.

These two observations establish that Axiom 2 for Boolean algebras holds here. We illustrate these two facts in the following circuits.

For each bit A, we note that if we call the opposite value of A, A', we find that A connected with A' through an *and* gate yields 0, whereas A connected with A' through an *or* gate yields 1; hence Axiom 3 for Boolean algebras is satisfied. These facts are illustrated in the following circuits.

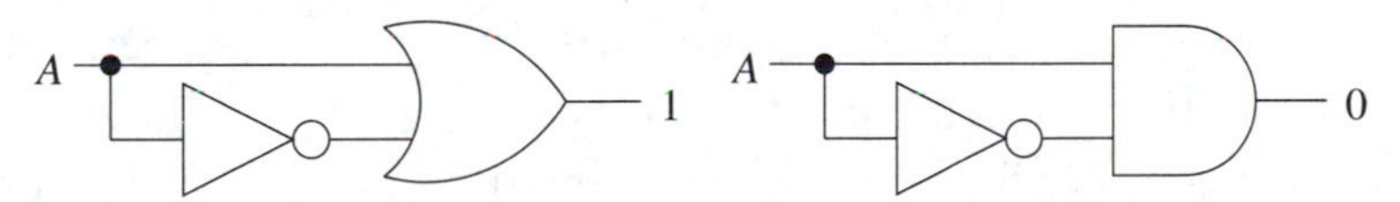

Axiom 4, distributivity, is left as an exercise, but it is straightforward to draw the appropriate circuits and trace through them to check that they function the same.

Having established that switching circuits (bits together with the *and*, *or*, and *not* gates) form a Boolean algebra, we are ready to have another look at the circuit on the bottom of page 49. We can now represent it using the notation of Boolean algebra and simplify the circuit from eight gates to two gates using the axioms and theorems we have already seen. This use of abstraction allows us to reason about circuits using a mathematical system that allows us to manipulate symbols without needing to think about the meanings of the symbols. A reduction of the expression for the circuit is given next. It should be straightforward to supply the reasons for each step.

$$\begin{aligned} A(B+C)+B(A+C')+CB &= AB+AC+BA+BC'+CB \\ &= BA+BA+AC+BC'+BC \\ &= BA+BA+AC+B(C'+C) \\ &= BA+AC+B1 \\ &= BA+B1+AC \\ &= B(A+1)+AC \end{aligned}$$

$$\begin{aligned} &= B1+AC \\ &= B+AC \end{aligned}$$

✓ Confirm, by constructing the appropriate truth tables, that $B+AC$ is indeed equivalent to the original expression. You will certainly agree it is much simpler!

Storing Numbers in a Digital Computer

Our study of switching circuits has given us a glimpse at design and simplification of circuitry, but we still have not addressed the question of how such circuitry can be used for more complex computer operations. We will now find out how to represent an integer in a computer and how to design circuitry that can add two integers.

Integers in digital computers are usually stored using the binary system. This system is analogous to the decimal system we use for human computation. All decimal numbers are represented in terms of 10 digits (0 through 9). The digits take on different meanings according to the position they hold. When we write 3649, we mean three thousands, six hundreds, four tens, and nine ones, or

$$3649 = 3 \cdot 10^3 + 6 \cdot 10^2 + 4 \cdot 10^1 + 9 \cdot 10^0.$$

Analogously, in the binary system, 100101 means one thirty-two, zero sixteens, zero eights, one four, zero twos, and one unit, or

$$100101 = 1 \cdot 2^5 + 0 \cdot 2^4 + 0 \cdot 2^3 + 1 \cdot 2^2 + 0 \cdot 2^1 + 1 \cdot 2^0.$$

We call this the *binary expansion* of the number. In decimal notation we would write this as 37, which we get by simply calculating the binary expansion of the binary number 100101.

Thus converting from binary to decimal is a straightforward computation. Similarly, we can convert a number written in any base to decimal. (Note that a number written in base b uses digits 0 through $b - 1$.) For example, we can convert the base 8 number 5036 to decimal:

$$5036 = 5 \cdot 8^3 + 0 \cdot 8^2 + 3 \cdot 8^1 + 6 \cdot 8^0 = 2590,$$

where 2590 is a decimal number. But we restrict our focus here to binary integers. The following table shows how we count in binary for the first few integers:

binary	decimal
0	0
1	1
10	2
11	3
100	4
101	5
110	6
111	7
1000	8
1001	9
1010	10
1011	11

✓ Represent the decimal integer 1000 using the binary system.

Now let's do some addition. Elementary school children are taught to add two integers by using an algorithm. Beginning with the unit position, add the digits. If the sum can be written with one digit, write that digit in the unit position of the answer. If the sum requires two digits, then the digit in the unit position of the sum should be written in the unit position of the answer and the second digit should be added to the two digits in the tens position of the numbers being added. This is called *carrying* the digit. This process continues from right to left until all the positions have been processed. (If one number has more digits than the other, we simply pad the shorter integer on the left with 0's.)

The same algorithm works for adding numbers in any base. For example, suppose we want to add two binary numbers, 10110 and 1111. We can write the numbers so that they are right justified, filling in on the left with zeros, if needed.

$$\begin{array}{r} 010110 \\ +\ 001111 \\ \hline 100101 \end{array}$$

Beginning at the right, we add 0 to 1. Since the sum takes only one digit, we fill it in. When we move to the 2^1 column, since the sum of 1 and 1 is 10, we write the 0 in the 2^1 column and *carry* the 1 into the next column to the left. We continue, arriving at the answer shown.

Circuitry to Add

Since digital computers use binary (base 2) representation, addition must be done in binary. Before designing circuits that do addition, we first consider how a computer might store integers. We will look here only at positive or unsigned integers. In most computers positive integers are stored in *words*. Each word has a fixed number of bits, usually 8, 16, or 32. The bits in the word represent the value of the integer in base 2.

For example, if we want to represent 99 (expressed in base

10) in a computer, we use 01100011, if the word size is 8, since

$$01100011 = 0{\cdot}2^7+1{\cdot}2^6+1{\cdot}2^5+0{\cdot}2^4+0{\cdot}2^3+0{\cdot}2^2+1{\cdot}2^1+1{\cdot}2^0 = 99.$$

If the word size were 16, then 8 0's would be used to fill out the left part of the word: 0000000001100011.

✓ Suppose a computer uses one byte (8 bits) to represent each integer. What happens if you add 99 to 98?

We've already seen that the same algorithm we learned for base 10 addition works for base 2 as well. Thus we will need to think of doing two operations for each position in an addition problem. First we get the sum. Second we find the carry value. The following table illustrates this process in binary. The x and y represent the binary digits in a given position. The s represents the sum bit and the c represents the carry-out bit.

x	y	c	s
0	0	0	0
0	1	0	1
1	0	0	1
1	1	1	0

This is the truth table for a circuit known as a "half adder." You probably have realized that this does not completely describe the process we need since there are really 3 bits to be added in each position: x and y as previously but also the carry-in bit. We'll consider this problem after we've designed the half adder.

We see that the carry-out bit is the value obtained by using the x and y as inputs to an *and* gate, something we have seen before. The sum value, however, requires us to look at one more kind of gate, one that is called the *xor gate*, standing for *exclusive or*. Given two inputs, the result of the *xor* is 1 if exactly one of the values of the inputs is 1. The output is 0 otherwise. (This is in contrast to our usual *or*, which is sometimes called the *inclusive or* since the result of *or*ing two values is true if either one or both is true.) The symbol for the *xor* operation in a Boolean expression is $\oplus$. So the expression for a *xor*ed with b would be $a \oplus b$. In terms of switching circuits, we express the *xor* as

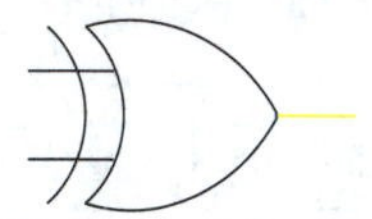

Thus the circuit for a half adder is

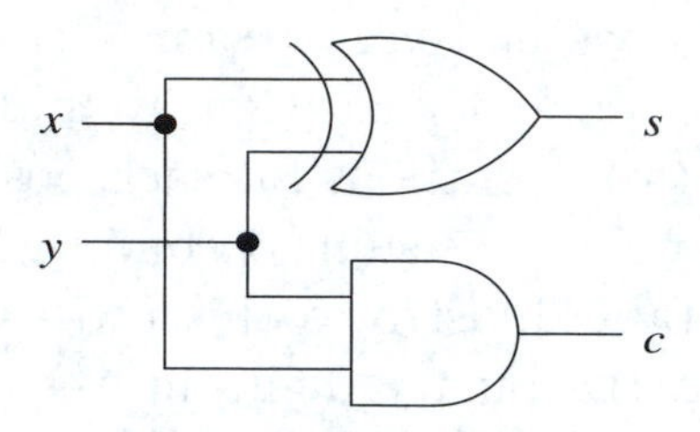

Note that instead of the *xor* gate we could have used the equivalent $x'y + xy'$, but that would have required more gates.

Let's expand and draw a full adder, one that has three inputs: a carry-in bit as well as bits x and y. There are still two outputs: carry-out and store. The truth table for the full adder is

c_{in}	x	y	c_{out}	s
0	0	0	0	0
0	0	1	0	1
0	1	0	0	1
0	1	1	1	0
1	0	0	0	1
1	0	1	1	0
1	1	0	1	0
1	1	1	1	1

We would like to derive Boolean expressions for c_{out} and s. Furthermore, we'd like these to be as simple as possible. What these expressions are is not as obvious as for the half adder. The method we'll use here is applicable to all truth tables. So you'll be able to easily get a Boolean expression for any truth table.

We'll derive a Boolean expression for c_{out} first. The first step is to write a Boolean expression for c_{out} in a standard form called *disjunctive normal form* (DNF). To do this, look at the 1's in the c_{out} column. For each 1, make a conjunction from the inputs that is true *only* for those inputs. That is, the truth table for that conjunction would be 0's everywhere except for a 1 on that line. The conjunction for the first 1 in the c_{out} column is clearly $c'_{\text{in}}xy$. The other conjunctions are $c_{\text{in}}x'y$, $c_{\text{in}}xy'$ and $c_{\text{in}}xy$.

Now make a disjunction of all these expressions; this will be the DNF for c_{out}:

$$c_{\text{out}} = c'_{\text{in}}xy + c_{\text{in}}x'y + c_{\text{in}}xy' + c_{\text{in}}xy.$$

It is important to note that the DNF for any truth table is unique. (You find it using the technique just described.) Also, once we have any Boolean expression for a truth table, we can then draw a switching circuit for it. Furthermore, since we can write a truth table for every switching circuit and every truth table has a DNF, we can draw a switching circuit for any Boolean function using only *and*, *or*, and *not* gates.

We could draw the circuit for c_{out} directly from its DNF, but, as we've seen, it's usually worthwhile to simplify the expression first. There are different directions we could go here. We'll use the following simplification which makes nice use of the exclusive or:

$$\begin{aligned} c_{\text{out}} &= c'_{\text{in}}xy + c_{\text{in}}x'y + c_{\text{in}}xy' + c_{\text{in}}xy \\ &= (c'_{\text{in}} + c_{\text{in}})xy + c_{\text{in}}(x'y + xy') \\ &= 1xy + c_{\text{in}}(x \oplus y) \\ &= xy + c_{\text{in}}(x \oplus y). \end{aligned}$$

Similarly, we can write the DNF for s and simplify:

$$\begin{aligned} s &= c'_{\text{in}}x'y + c'_{\text{in}}xy' + c_{\text{in}}x'y' + c_{\text{in}}xy \\ &= c'_{\text{in}}(x'y + xy') + c_{\text{in}}(x'y' + xy) \\ &= c'_{\text{in}}(x \oplus y) + c_{\text{in}}(x \oplus y)' \\ &= c_{\text{in}} \oplus (x \oplus y). \end{aligned}$$

Note that there are other expressions possible for s and c_{out} but these are convenient since one gate, $x \oplus y$, is used in both circuits. Thus the circuit for a full adder is

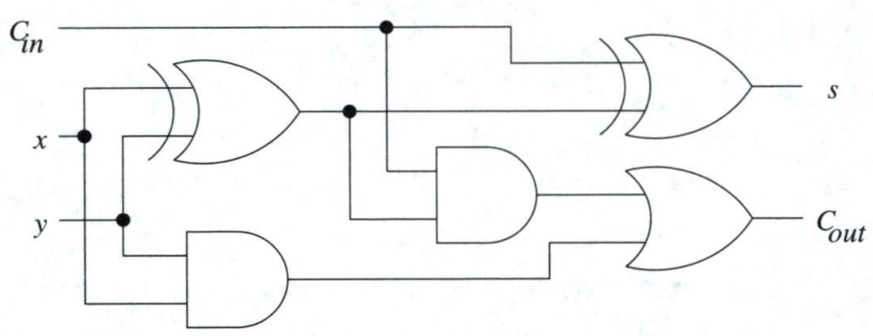

We typically use simpler circuits to build up more complex circuits. When doing this, the simple circuits are treated as "black boxes" where the inputs and outputs are labeled, but the internal circuitry is not given. This simplifies the design process. For example, the black box units for the half adder and the full adder might be drawn as

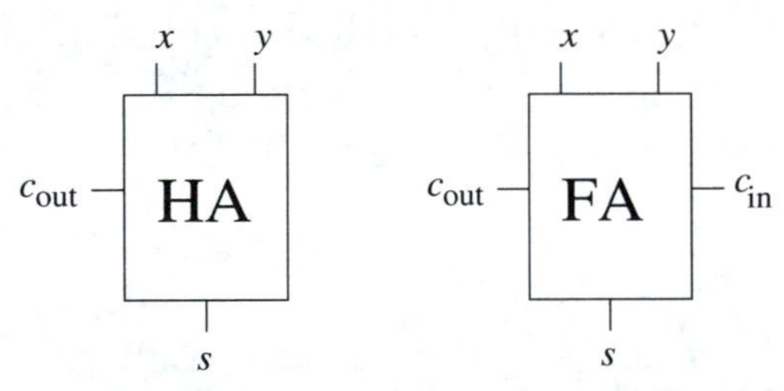

We can now string together four full adders to make a 4-bit adder by connecting the c_{out} of a full adder to the c_{in} of the full adder to its immediate left. We connect the carry-in to the rightmost adder to 0. The reason we don't use a half

adder for this rightmost bit is that then we can use this 4-bit adder as a building block for larger adders. We'll leave this as an exercise.

Exercises

1. For propositional logic, prove the following, if you have not already done so:
   a. $\wedge$ and $\vee$ are commutative and associative.
   b. $\vee$ distributes over $\wedge$.
   c. T is the identity for $\wedge$ and F is the identity for $\vee$.
2. a. Write $p \Rightarrow q$ in DNF.
   b. Simplify your expression (use only $\vee$, $\wedge$, and $\neg$).
3. Construct a truth table for $\neg(p \vee \neg q) \Rightarrow \neg p$.
4. Construct a truth table for $(p \wedge q) \Rightarrow p$.
5. Write the negation and simplify $(q \vee r) \wedge (\neg q \vee r)$.
6. Write the negation and simplify $p \vee q \vee (\neg p \wedge \neg q \wedge r)$.
7. The following are known as the *absorption laws* for logic: $p \vee (p \wedge q) = p$ and $p \wedge (p \vee q) = p$. Prove these using truth tables.
   How would you write the corresponding statements for Boolean algebras?
8. (*) You can prove the absorption laws for *all* Boolean algebras using just stuff you know about Boolean algebras. But there's a little trick. (There always is.) Here's the first step in showing that $p + (p * q) = p$: $p + (p * q) = (p * 1) + (p * q) = \cdots$. Now continue reducing until you get p.
   Use a similar trick to show the other absorption law for Boolean algebras.
9. Simplify: $AB + (B'C)' + (D + C')$. Draw the simplified circuit.
10. It turns out you can construct the equivalent to an *or* gate using just *and* and *not* gates. Do so. (*Hint*: Think DeMorgan's laws.)

11. Likewise, you can construct the equivalent to an *and* gate using just *or* and *not* gates. Do so.
12. What is the largest (unsigned) integer you can store in 4 bits? In 8 bits? in 16 bits? In n bits? (Give your answers in decimal.)
13. Add these binary integers: 11010 + 1011, 1011 + 110, 1111110111 + 1001.
14. Convert these decimals to binary: 90, 52, 41.
15. Convert from binary to decimal: 111101, 100010, 111000000.
16. If an integer is written in binary, how can you easily tell it is even? If it is odd?
17. When doing arithmetic in the base 10 system, multiplying and dividing by 10 or powers of 10 can be done easily by moving the decimal point. How can you do multiplying and dividing by 2 and powers of 2 in the binary system?

* 18. Suppose you want to represent real numbers in binary form. How would you represent one half? *Hint*: Since on the right of the binary point, each position will stand for a negative power of 2, the first position stands for 2^{-1}, the second position for 2^{-2}, and so on.

* 19. Given the base 10 numeral 234.75, represent the number it stands for in base 2.

20. Add the binary numbers: 110101.101 + 1100011.011.

* 21. If a and b are integers, what is $(a+b)^2$? $(a+b)^3$? $(a+b)^n$? If a and b are elements of a Boolean algebra, what is $(a+b)^2$? $(a+b)^3$? $(a+b)^n$? (Here, $(a+b)^2$ is shorthand for $(a+b)*(a+b)$. The other exponents stand for similar expressions.)

* 22. Show that if B is a Boolean algebra, then there can be no element $a \in B$ where $a \neq 0$ and $a \neq 1$ such that $a' = a$. (That is, a can't be self-dualing.)

* 23. Show that there is no Boolean algebra with exactly three elements. (*Hint*: Use the previous exercise.)

* 24. Can you find a Boolean algebra with four elements? With five elements? With six elements? For an extra challenge, can you generalize?

25. You showed that the collection of subsets of any given finite set can form a Boolean algebra with appropriately chosen operations. Is the same true of an infinite set?

26. Suppose p is a proposition. What is the dual of p? (Note that for your choice to work properly, Axiom 3 must be satisfied.) Verify that Axiom 3 is indeed satisfied. Now show that the well-formed expressions of propositional logic form a Boolean algebra.

27. Write the duals of all the theorems in this chapter.

* 28. We noted that the dual of a Boolean algebra theorem is one where the +'s and *'s are exchanged and the 0's and 1's are exchanged in the original theorem. The *principle of duality* says that the dual of any Boolean algebra theorem is also true. Prove that the four axioms of Boolean algebra are self-dualing. That is, the dual of each axiom is itself. Why does the principle of duality follow from this?

29. Consider the collection of all integers with the usual elements 0 and 1, together with the usual addition and multiplication. We've seen that this is not a Boolean algebra since $+$ does not distribute over $*$. Which of the other axioms for Boolean algebras do the integers satisfy and which do they not satisfy?

30. Let $B = \{0, 1\}$, with $+$ defined to be the usual addition modulo 2 and $*$ the usual multiplication. Is B a Boolean algebra?

* 31. Prove Theorems 3 through 8 on Boolean algebras.

32. If a, b, and c are elements in some Boolean algebra, show that the following four expressions are equivalent:

$$(a+b)*(a'+c)*(b+c) \qquad a*c+a'*b+b*c$$
$$(a+b)*(a'+c) \qquad a*c+a'*b$$

33. Show that $a * b + b * c + c * a = (a + b) * (b + c) * (c + a)$, where a, b, and c are elements of a Boolean algebra.

* 34. If a, b and c are elements of some Boolean algebra, and if both $a * b = a * c$ and $a + b = a + c$, then show that $b = c$. (*Hint*: Use the absorption laws from a previous exercise.)

* 35. If a, b, and c are elements of some Boolean algebra, then show that if $a * x = b * x$ and $a * x' = b * x'$ for all x in the Boolean algebra, then $a = b$.

* 36. If a, b, and c are elements of some Boolean algebra, define $a \leq b$ if and only if $a * b' = 0$. Show that if $a \leq b$, then $a + b * c = b * (a + c)$ for all c.

* 37. For a, b, and c in some Boolean algebra, show that
   a. If $a \leq b$ and $b \leq c$, then $a \leq c$.
   b. If $a \leq b$ and $a \leq c$, then $a \leq bc$.
   c. If $a \leq b$, then $a \leq b + c$ for all c.
   d. $a \leq b$ if and only if $b' \leq a'$.

* 38. For a, b, and c in some Boolean algebra, show that $a = b$ if and only if $a * b' + a' * b = 0$.

39. Let $S = \{1, 2 \ldots, n\}$ for some positive integer n. On S define $x + y = \max\{x, y\}$ and $x * y = \min\{x, y\}$. Can you find a way to make S a Boolean algebra with these two operations?

* 40. Suppose a and b are elements of a Boolean algebra. Show that the following are equivalent: $ab = a$, $a + b = b$, $a' + b = 1$, $ab' = 0$.

41. Let S be the set of positive divisors of 110 (i.e., $S = \{1, 2, 5, 10, 11, 22, 55, 110\}$). Show that S together with the operations gcd (greatest common divisor) and lcm (least common multiple) forms a Boolean algebra. You will need to figure out what the dual of each element is and what the zero element and the one element should be.

42. Let S be the set of positive divisors of 18 (i.e., $S = \{1, 2, 3, 6, 9, 18\}$). Show that this set together with gcd and lcm does not form a Boolean algebra.

* 43. Based on the previous two exercises, make a conjecture as to what sets of divisors do make Boolean algebras when the operations are gcd and lcm.

* 44. A set is called "cofinite" if its complement is finite. Let U be the set of all finite and cofinite subsets of the natural numbers. Prove that the subsets of U together with the operations union, intersection, and complement, forms a Boolean algebra.

45. Construct circuits for these expressions: (i) $AB'+CB'+A$, (ii) $(A+C')(B+C')A$.

46. Simplify the expressions in Exercise 45 and draw the circuits.

47. Draw a circuit with four inputs whose output is their sum (in binary). Note that the possible outputs are 0, 1, 10, 11, and 100. Thus you need how many bits for output here? (Recall that each output line means a new circuit.) You may use full or half adders in your circuit, if you wish.

48. Draw a circuit with three inputs, call them A, B, and C, whose output is 1 if A equals the sum of B and C.

49. Draw a circuit with three inputs whose output is 1 when an odd number of inputs are 1. (Try to use XOR gates in your circuit to make it simple.)

50. Draw a switching circuit for the expression $A(B' + A'C)$. Same question for $(B + A')(AC' + B'C)$.

51. Simplify the two expressions in the previous exercise and redraw the circuits.

52. Build a 4-bit adder from four full adders. The inputs for this circuit will be the four bits for one 4-bit summand, A_0, A_1, A_2, A_3; another four bits for the other summand, B_0, B_1, B_2, B_3; and the carry-in for the right-most bit, c_{in}. Note that c_{in} will be wired to 0 when using this as a stand alone 4-bit adder. What should the outputs be? What would the black box diagram for this 4-bit adder look like? Using your black box diagram, design an 8-bit adder.

53. Give a Boolean expression for each circuit shown.

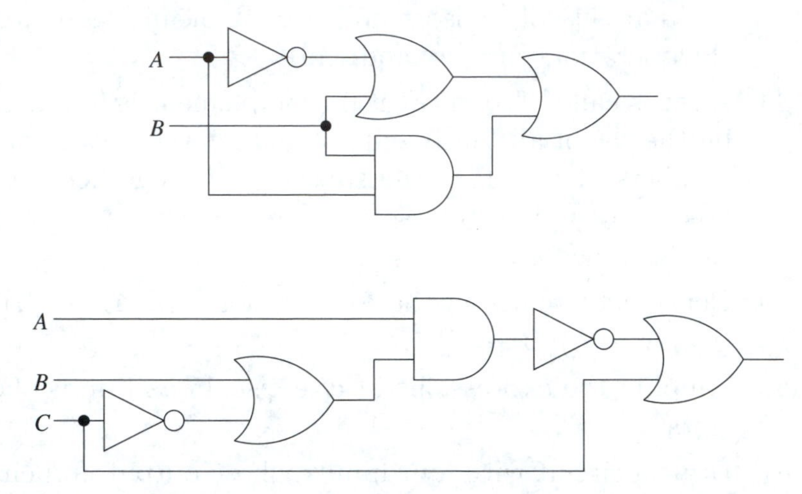

54. Simplify the Boolean expressions you got in the previous exercise and draw the circuits you get. Give the truth tables for both the original circuit and the simplified ones and check that they are indeed equivalent.

* 55. Suppose we call two circuits *equivalent* if they have the same truth table. We have seen that for any one circuit, there are many circuits equivalent to it. Call the collection of all circuits that are equivalent an *equivalence class* of circuits. Now consider all circuits with two inputs and one output. How many equivalence classes are there for these circuits? (This is the same as the question "How many different truth tables are there with two inputs?") Answer the same question for three-input circuits. Answer the same question for circuits with n inputs.

Programming Problems

1. Write a program that takes as input a number represented in base 2 and produces as output the same number represented in base 10. Write a similar program for base b.

2. Write a program that takes as input two numbers, x, y, and a base, b, and produces as output the sum of x and y

(in base b, of course).

3. Write a program that permits users to do arithmetic in any base between 2 and 10. They should be able to add, subtract, multiply, and divide. A user can select a base and then perform whatever arithmetic operation is required.

* 4. Write a program that, given a logic expression for propositional logic, prints out the corresponding truth table. For a simpler version, logic expressions may be restricted to either the negation of a single proposition or two propositions connected with a single connective.

5. Write a program to exhibit all possible truth tables for the propositional logic when only two propositions are involved.

* 6. Write a program that serves as a binary adder for very large integers. (There should be no limit to the number of bits stored here.)

Chapter 4 Natural Numbers and Induction

All mathematics begins with counting. This is the process of putting the set of objects to be counted in one-to-one correspondence with the first several *natural numbers* (or *counting numbers*):

$$1, 2, 3, 4, 5, \cdots.$$

We denote this infinite set by $\mathbb{N}$. We should note here that there is not general agreement on the definition of $\mathbb{N}$; some mathematicians also include 0 in $\mathbb{N}$. You will encounter no great trouble one way or the other; which definition you use is merely a matter of convenience. Indeed, as we shall see, both of these sets share the fundamental properties important to us here. Many of the uses of natural numbers in computer science are obvious, but their most powerful property, the principle of mathematical induction (or simply induction), is the one we'll emphasize in this chapter. In the next chapter, we'll focus on the arithmetic of $\mathbb{N}$.

The importance of induction cannot be overstated. Many important facts about natural numbers are proved using this technique, and we will do that in this chapter. But beyond the mathematical proofs, induction offers a way of thinking about processes—processes that you will want to program on a computer. Using induction to write programs goes by another name: recursion. This is a topic of a later chapter. Being able to think recursively is an important step in your development as a computer scientist. In this chapter we will discuss some properties of natural numbers informally and then finally give a formal set of axioms.

Well-ordering and Mathematical Induction

A fundamental property of $\mathbb{N}$ is *well-ordering*, a property that we state formally below, and that we shall accept as an axiom about $\mathbb{N}$. An axiom for any system is a statement we accept as true without proof. In order to reason about a system—in this case the natural numbers—we must start by assuming something about it. These baseline assumptions should be reasonable and "obviously true," although there is frequently more than meets the eye. While we won't be strictly rigorous here, we've included one set of axioms for $\mathbb{N}$ at the end of this chapter. These are called the Peano axioms and they seem very primitive indeed. From them we could prove all important provable facts about $\mathbb{N}$. Note, however, that there are other sets of axioms for $\mathbb{N}$ that are equally as valid. But our purpose here is to convince you that mathematical induction is valid. We could simply ask you to accept it as a reasonable assumption—that is, as an axiom—but the well-ordering principle seems to be more intuitive as an axiom and so we'll start by asking you to accept it.

The Well-ordering Principle. *Every nonempty subset of* $\mathbb{N}$ *has a least element.*

For any subset of $\mathbb{N}$ that we might specify by actually listing the elements, this is obvious, but the principle applies even to sets that are more indirectly defined. For example, consider the set of all natural numbers expressible as $12x + 28y$, where x and y are allowed to be any integers. The extent of this set is not evident from the definition. Yet the well-ordering principle applies, and since this set is nonempty (which is important to note) there is a smallest natural number expressible in this way. (Of course, finding the value of that smallest number is another matter. Note that the well-ordering principle guarantees existence of a smallest number but says absolutely nothing about how to calculate it.)

Suppose we wish to apply the well-ordering principle to a particular subset X of $\mathbb{N}$. We may then consider a sequence of yes/no questions of the following form:

$$\text{Is } 1 \in X?$$
$$\text{Is } 2 \in X?$$
$$\vdots$$

Because X is nonempty, sooner or later one of these questions must be answered yes. The first such occurrence gives the least element of X. Of course, it might not be easy to answer such questions in practice. But nevertheless, the well-ordering principle asserts the existence of this least element, without identifying it explicitly.

The well-ordering principle allows us to prove one of the most powerful techniques of proof that you will encounter. (We'll save the proof until later in the chapter.) This is the *principle of mathematical induction*:

Principle of Mathematical Induction. *Suppose X is a subset of $\mathbb{N}$ that satisfies the following two criteria:*

(1) $1 \in X$, *and*

(2) *For all $n > 1$, $k \in X$ for all $1 \leq k < n$ implies $n \in X$.*

Then $X = \mathbb{N}$.

The principle of mathematical induction is used to prove that certain sets X equal the entire set $\mathbb{N}$. In practice, the set X will usually be "the set of all natural numbers with property such-and-such." For example, X could be all natural numbers n that satisfy the equation $1 + 3 + \cdots + (2n - 1) = n^2$ (which we will prove soon). To apply it we must check two things:

(1) the "base case": that the least element of $\mathbb{N}$ belongs to X, and

(2) the "bootstrap" or the "induction step": a general statement that asserts that a natural number belongs to X whenever all its predecessors do.

You should find the principle of mathematical induction plausible. First, the base case establishes that $1 \in X$. Then successively applying the bootstrap allows you to conclude that

$$2 \in X,\ 3 \in X,\ 4 \in X,\ \cdots.$$

Thus *every* natural number is in X. That is, $X = \mathbb{N}$.

Let's look at the induction step more closely. In the induction step we assume that $n > 1$, since the case when $n = 1$ is handled in the base case. Notice that the induction step is an implication. Indeed, we could write it in the form IF $k \in X$ for every $1 \leq k < n$ THEN $n \in X$. The clause "$k \in X$ for every $1 \leq k < n$" is called the *induction hypothesis*. This statement you assume to be true when showing the induction step. Under this assumption you need to show that it follows that $n \in X$. If you successfully do this and you establish the truth of the base case, then mathematical induction says that $X = \mathbb{N}$.

When checking the bootstrap, we assume that all predecessors of n belong to X, and must infer that n belongs to X. In practice we often need only that certain predecessors of n belong to X. Specifically, many times we will need only that $n - 1$ belongs to X. Indeed, the form of induction you may have used before probably assumed only that $n - 1$ was in X, instead of all $1 \leq k < n$. It turns out that the version you learned before and the version we will be using are equivalent, although they don't appear to be at first glance. We will find the version given here of more use.

Before proving the principle of mathematical induction itself, let us look at some examples of its use. As beginners in using mathematical induction, we'll carefully establish each part necessary for the proof. Specifically, we'll (1) explicitly declare what the set X is, (2) state the base case, (3) prove the base case, (4) declare explicitly the induction hypothesis, (5) declare explicitly what we wish to show in the induction step, and (6) prove that what we've stated in part 5 follows from the induction hypothesis.

For our first example, let's show that the sum of the first n odd integers is n^2. That is,

$$1 + 3 + 5 + \cdots + (2n - 1) = n^2, \text{ for } n \geq 1.$$

Proof by Induction: Here the set X are those $n \geq 1$ for which the equation $1 + 3 + 5 + \cdots + (2n - 1) = n^2$ holds.

First we note that the base case is $n = 1$ here. The left of the equation has one term, $2 \cdot 1 - 1$. But $2 \cdot 1 - 1 = 1^2$, and so our formula certainly holds for $n = 1$, thus proving the base case that $1 \in X$.

We now do the induction step and so assume that $n > 1$. The induction hypothesis is that $k \in X$ for $1 \leq k < n$. Using the induction hypothesis, we must show that $n \in X$. Explicitly, we are assuming that if $1 \leq k < n$, then

$$1 + 3 + 5 + \cdots + (2k - 1) = k^2.$$

Now we need to show that $n \in X$, assuming our induction hypothesis is true. That is, we need to show that

$$1 + 3 + 5 + \cdots + (2n - 1) = n^2.$$

But by our induction hypothesis, by putting $k = n - 1$, we have

$$1 + 3 + 5 + \cdots + (2(n - 1) - 1) = (n - 1)^2.$$

Thus,

$$\begin{aligned} 1 + 3 + 5 + \cdots + (2(n - 1) - 1) + (2n - 1) &= \\ (n - 1)^2 + (2n - 1) &= \\ n^2 - 2n + 1 + (2n - 1) &= n^2, \end{aligned}$$

which shows that the formula holds for n. That is, we've shown $n \in X$. Thus, by the principle of mathematical induction, $X = \mathbb{N}$; that is, the formula holds for *all* $n \geq 1$. //

Our next two examples prove something about full binary trees. We need to describe these first. A *binary tree* consists of a set of nodes (just points to us), connected by edges. The nodes are arranged in a hierarchy. The node at the top is the *root node*, and the nodes directly beneath, which are connected to it by edges, are the node's *children*. These nodes can in turn have children, which in turn can have children, and so on. What makes this a *binary* tree is that any node can have at most two children. The following are three binary trees.

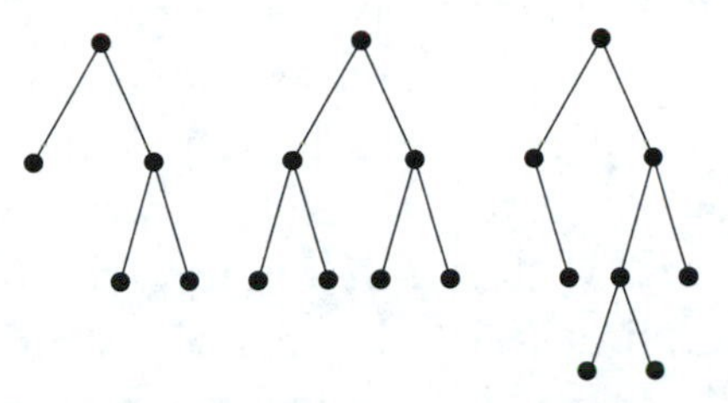

Notice that every node in each of these three trees has either zero, one, or two children. Now the nodes are at various *depths*. The root node is at depth 1. The children of the root node are at depth 2 and so on. The *depth of the tree* is the maximum depth of any node in the tree. Thus the first two binary trees are of depth 3 while the third one is of depth 4. A *full binary tree* is one where every node has two children, except those nodes at the deepest level. So the middle tree in the preceding diagram is a full binary tree of depth 3. Notice that a full binary tree of depth d has the maximum number of nodes of all binary trees of depth d.

We'll now prove a couple of facts about full binary trees. We will first show that the number of nodes at level n of a full binary tree of depth n is 2^{n-1}.

> ✓ Before proceeding, draw the full binary trees of depths 1, 2, 3, and 4 and check that indeed the preceding statement is true.

Proof by Induction: Let X be the set of positive integers n for which the preceding statement is true; that is, X is those positive integers n where the number of nodes at level n of a full binary tree of depth n is 2^{n-1}.

The base case is when $n = 1$. We need to show that the number of nodes at depth 1 in a full binary tree of depth 1 is $2^{1-1} = 1$. But, of course, a full binary tree of depth one consists only of its root node! (In fact this is true of any binary tree of depth 1.) So we see that the base case is true.

Now let's suppose $n > 1$. Our induction hypothesis is that for $1 \leq k < n$, a full binary tree of depth k has 2^{k-1} nodes at depth k. We wish to show that this implies that a full binary tree of depth n has 2^{n-1} nodes at depth n. So let's take a full binary tree of depth n. If we strip away all the nodes at depth n, we are left with a full binary tree of depth $n - 1$. By our induction hypothesis with $k = n - 1$, we are assuming that there are $2^{n-1-1} = 2^{n-2}$ nodes at depth $n - 1$. But each of these 2^{n-2} nodes has two children, and these children are exactly all the nodes at level n of our original full binary tree. Thus there are $2 \cdot 2^{n-2} = 2^{n-2+1} = 2^{n-1}$ nodes at depth n, which is what we wanted to show. Therefore, we've established that $n \in X$ and so the induction step has been proved.

Thus mathematical induction says that our statement holds for all positive integers n. //

Now we'll use induction, and the last fact we proved, to show that a full binary tree of depth n has a total of $2^n - 1$ nodes. Before doing so, let's look at the first few small examples just to see if this is correct for those. This is a good idea in general since besides verifying the statement for small examples, it frequently suggests how to progress in proving the statement.

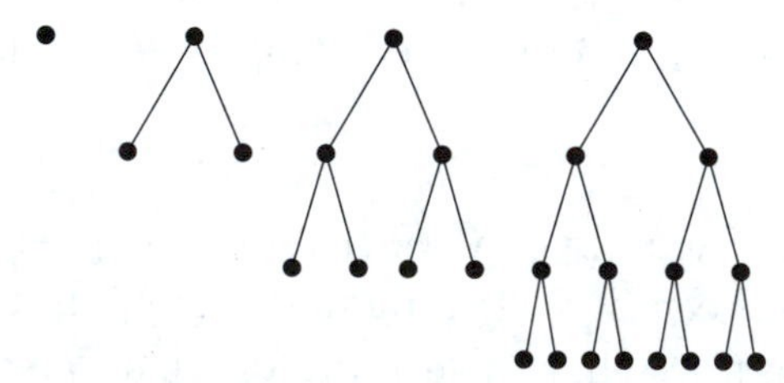

As we can easily see, these trees have 1, 3, 7, and 15 nodes and are of depths 1, 2, 3, and 4. We see that the formula $2^n - 1$ for the number of nodes in a full binary tree of depth n is true at least for these four small cases. But we wish to prove it in general. The last proof suggests that when trying to prove the case for depth n, we can look at the next smaller instance, which is of depth $n-1$. If we look at the case when $n = 4$, we see that stripping off the nodes at depth 4 leaves a full binary tree of depth 3. We know how many nodes are in that tree. If we just add those nodes at depth 4, we will get the total number of nodes. This way of thinking about this problem, as a smaller instance of the problem with some additional nodes, is the inductive way of thinking. The proof follows along exactly those lines.

Proof by Induction: Let X be the set of positive integers n where the full binary trees of depth n have $2^n - 1$ nodes. The base case is when $n = 1$. That is, the full binary tree of depth one has $2^1 - 1$ nodes. But this is 1, which is clearly the number of nodes in a full binary tree of depth 1.

Now we do the induction step and so assume that $n > 1$. Our induction hypothesis is that for $1 \leq k < n$, a full binary tree of depth k has a total of $2^k - 1$ nodes. Assuming this, we must show that a full binary tree of depth n has $2^n - 1$ total nodes.

So let's take a full binary tree of depth n. As in the last proof, if we strip away all the nodes at depth n, we are left with a full binary tree of depth $n-1$. If we let $k = n-1$, our induction hypothesis says that we have a total of $2^{n-1}-1$ nodes in our full binary tree of depth $n-1$. Now, by what we last proved, there are 2^{n-1} nodes at depth n. If we add this to the number of nodes for a full binary tree of depth $n-1$, we'll have the number of nodes of a full binary tree of depth n. This makes a total of $2^{n-1} - 1 + 2^{n-1} = 2 \cdot 2^{n-1} - 1 = 2^{n-1+1} - 1 = 2^n - 1$ nodes, as desired. Thus $n \in X$ and we've established the induction step.

Therefore, by induction, there are a total of $2^n - 1$ nodes on a full binary tree of depth n, for all $n \geq 1$. //

Our next example is a more geometric problem and illustrates the surprising versatility of the induction method. Here, we are concerned with *tiling* a figure. The tile we'll be using is an L-shaped piece as follows:

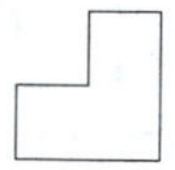

You can think of this tile as three boxes glued together, and we can rotate it in any manner when using it to tile an area. The area we're interested in is a large square made up of 2^n-by-2^n boxes, where one of the boxes has been removed. The L-shaped tile we'll use is made of three boxes. The question is can we tile (that is, completely cover with no overlaps and no pieces "sticking out") all such squares using only the L-shaped tiles. Note that the restriction to the dimensions is that they must be powers of 2. The follwoing are two examples, one of a 4-by-4 square with one box removed and one of an 8-by-8 square with one box removed. Note that there may be more than one way to tile this object.

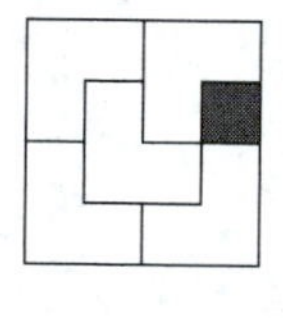

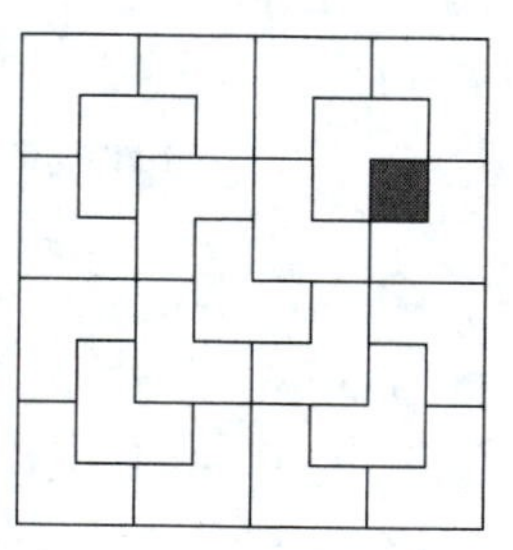

Before proceeding, try to tile an 8-by-8 square with a different box removed. The statement claims that this can be done regardless of which box you remove.

Let's restate our proposition: For $n \geq 1$, every square of 2^n-by-2^n boxes with one box removed, can be tiled using L-shaped tiles.

> ✓ Before proceeding, state the base case clearly and show that it is true.

Proof by Induction: Let X be the set of positive integers n for which every square of 2^n-by-2^n boxes with one box removed, can be tiled using L-shaped tiles. We first check that $1 \in X$. That is, that every 2-by-2 square with one box removed can be tiled with L-shaped tiles. But every such arrangement is a rotation of the following figure. This is easily seen to be tiled by one L-shaped tile.

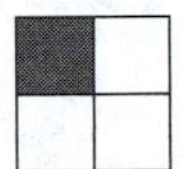

Now let's do the induction step. We assume $n > 1$. Our induction hypothesis is that for $1 \leq k < n$, every 2^k-by-2^k square with one box removed can be tiled using L-shaped tiles. We need to show that this implies that every square of 2^n-by-2^n boxes, with one box removed, can be tiled using L-shaped tiles.

To prove this, we start with a 2^n-by-2^n square with one box removed. We subdivide the large square into four squares, each of dimension 2^{n-1}-by-2^{n-1}. The missing box is in one of these four smaller squares. Now take where the three 2^{n-1}-by-2^{n-1} squares with no missing box join and place an L-shaped tile so that it covers one box in each of the three 2^{n-1}-by-2^{n-1} squares. This step is illustrated as follows:

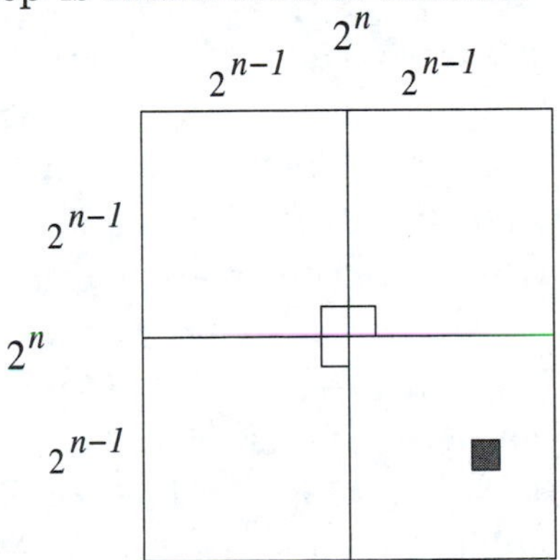

Note that now all four 2^{n-1}-by-2^{n-1} squares have one little square removed, either by the L-shaped tile or the original missing little square. So, letting $k = n - 1$, our induction hypothesis says we can tile each of these using L-shaped tiles. Those tiles, along with the one L-shaped tile we placed in the middle, will tile the original 2^n-by-2^n square. And so we have shown $n \in X$. Thus mathematical induction implies we can do this for any such squares. //

As a bonus, the proof of the last problem actually gives you a method of doing the tiling. We leave it as an exercise to do this.

As our next example, let's show that a finite set with n elements has exactly 2^n subsets. This important fact is used frequently.

> ✓ Before proceeding, state the base case clearly and show that it is true.

Proof by Induction: Let X be the set of those positive integers for which a set with n elements has exactly 2^n subsets. We first check that $1 \in X$, which is the base case. But a set with one element has itself and the empty set as subsets. This is $2 = 2^1$ subsets, as required.

Now let's do the induction step. So, suppose that $n > 1$. Our induction hypothesis is that $k \in X$ for all $1 \leq k < n$; that is, a set with k elements has exactly 2^k subsets, for all k where $1 \leq k < n$. We must prove that $n \in X$. Suppose then that S is a set with n elements; we must show that S has 2^n subsets. Because S has at least one element, choose one of them and call it s. Now every subset of S either contains s or it doesn't. Those subsets that don't contain s are precisely the subsets of $S \backslash \{s\} = \{x \in S : x \neq s\}$. But this set has $n - 1$ elements, and so by our assumption that $n - 1 \in X$, we know that $S \backslash \{s\}$ has 2^{n-1} subsets. Now those subsets of S that *do* contain s are of the form $A \cup \{s\}$, where A is a subset of $S \backslash \{s\}$. There are also 2^{n-1} of these subsets. Thus, there are $2^{n-1} + 2^{n-1} = 2^n$

subsets of S altogether. In other words, $n \in X$. Thus, by induction, $X = \mathbb{N}$; that is, any finite set with n elements has exactly 2^n subsets. //

Note that the preceding theorem also holds when $n = 0$, since we are talking about the empty set that has one subset. (What is it?) Note that the set $\{0, 1, 2, 3, \cdots\}$ is also well-ordered (we've only added 0). Thus Mathematical Induction applies to this set as well. (This is what we referred to at the beginning of the chapter when saying that $\mathbb{N}$ and $\mathbb{N} \cup \{0\}$ share the same fundamental properties.) So the induction of the previous theorem could have started at $n = 0$ (the base case) which is considering the subsets of the empty set. Check that the induction step still works as written. The sets $\{2, 3, 4, \cdots\}$, $\{3, 4, 5, \cdots\}$, and so on are also well-ordered. Hence induction can be applied to these sets, too. To use induction here means only that our base case is when n equals the smallest element of the set in question. With a base case of $n = 2$, say, remember that we would then have proved that $X = \mathbb{N} \backslash \{1\}$.

We now look at a second form of induction. This form allows us to handle situations with a slightly more complicated base case.

Alternate Version of the Principle of Mathematical Induction.

Suppose $n_0 \in \mathbb{N}$ and X is a subset of $\mathbb{N}$ that satisfies the following two criteria:

(1) $1, 2, \ldots, n_0 \in X$, *and*

(2) *For all $n > n_0$, $k \in X$ for all $1 \leq k < n$ implies $n \in X$.*

Then $X = \mathbb{N}$.

Note that the base case has been expanded to not one but many (although finite) instances. The induction step then starts by considering n *greater than* the largest instance handled in the base case. As before, we can revise this to start at any number, since the sets involved are still well-ordered;

for instance, the sets $\{0, 1, 2 \ldots\}$ and $\{6, 7, 8, \ldots\}$. If we were interested in proving something for all $n \geq 6$ using this version, our two steps in an inductive proof would be to show, for some $n_0 > 6$, that

(1) $6, 7, \ldots, n_0 \in X$, and

(2) for all $n > n_0$, $k \in X$ for all $6 \leq k < n$ implies $n \in X$.

If we do this then the alternate version of induction allows us to conclude that $\{6, 7, 8, \ldots\} = X$.

Let's do an example. Let's show that every integer larger than 5 can be written as a sum of 3's and 4's. This problem is a little different from the previous two because the base case is more complex; this problem will nicely illustrate the alternate version of induction. We'll give the proof first, and then you'll see why the base case must be what it is.

Proof by Induction: Let X be the set of natural numbers that can be written as a sum of 3's and 4's. The base case here is to show the proposition holds for 6, 7, and 8. But you can see that 6, 7, and 8 can all be written as sums of 3's and 4's. (We'll leave it to you to fill in this detail.) Thus, $6, 7, 8 \in X$. Now for the induction step, let $n > 8$ and assume that if $6 \leq k < n$, then $k \in X$; that is, assume k can be written as a sum of 3's and 4s'. We want to show that it follows that n can also be written as a sum of 3's and 4's. But if $n > 8$, then $n - 3 \geq 6$ and so, by our assumption, $n-3$ can be written as a sum of 3's and 4's. But then n can also since we just add a 3 to the 3's and 4's that sum to $n - 3$. So $n \in X$. Thus induction implies that all integers greater than or equal to 6 can be written as a sum of 3's and 4's. //

Note the base case here had to account explicitly for the three smallest cases since our induction step had to dip down to $n - 3$.

Students new to induction often feel that in verifying the bootstrap they are assuming exactly what they are required to prove. This feeling arises from a misunderstanding of the fact that the bootstrap step is an *implication*: that is, a statement of the form $p \Rightarrow q$. To prove such a statement, we must assume p and then derive q.

Well-ordering Implies Mathematical Induction

As promised, we now prove the principle of mathematical induction, using the well-ordering principle.

The Well-ordering Principle implies the Principle of Mathematical Induction.

Proof: Suppose that X is a subset of $\mathbb{N}$ satisfying both criteria (1) and (2). Our strategy for showing that $X = \mathbb{N}$ is "proof by contradiction."

In this case we assume that X is a *proper* subset of $\mathbb{N}$, and so $Y = \mathbb{N} \backslash X$ is a non-empty subset of $\mathbb{N}$. By the well-ordering principle, Y possesses a least element m. Clearly, $m \neq 1$ by (1). All natural numbers $k < m$ belong to X, because m is the *least* element of Y. However, by (2) we conclude that $m \in X$. But now we have concluded that $m \in X$ and $m \notin X$; this is clearly a contradiction. Our assumption that X is a proper subset of $\mathbb{N}$ must have been false. Hence, $X = \mathbb{N}$. //

Amazingly, perhaps, if you assume the principle of mathematical induction, then you can prove the well-ordering principle. (We won't do that here, but you're welcome to try your hand at the proof.) Thus, the two ideas are equivalent! It is simply a matter of which of the two ideas you are going to accept as an axiom. It is largely a matter of taste, but it boils down to which of the two ideas seems most obvious and so is the easier to accept as an axiom. We picked well-ordering. In the next section, we present the Peano axioms, a set of axioms from which you can rigorously deduce all the facts of the natural numbers.

The Peano Axioms

Many of the properties of the natural numbers we've used are taken for granted. It is possible to be rigorous in establishing some basic properties of the natural numbers as opposed to relying on our intuition. We have taken this approach here because it is induction that we want to focus on and not the foundations of mathematics. But you should be aware that all properties of natural numbers, such as associativity and commutativity of addition and even the definition of addition itself, can be established from a few basic axioms. We give you a little taste of that in this section. By making our assumptions clear and our proofs careful, we will be able to accept with confidence the truth of statements about the natural numbers, even if the statements themselves are not obviously true.

The first extended example of an axiomatic approach to mathematics appears in *The Elements* of Euclid, who was a Greek mathematician living circa 300 *B.C.* In his book he developed much of ordinary plane geometry by means of a careful logical string of theorems, based on only five axioms and some definitions. The logical structure of Euclid's book is a model of mathematical economy and elegance. So much mathematics is inferred from so few underlying assumptions!

Note, of course, that we must accept *some* statements without proof (and we call these statements *axioms*)—for otherwise we'd be led into circular reasoning or an infinite regress.

One cost of the axiomatic method is that we must sometimes prove a statement that already seems obvious. But if we are to be true to the axiomatic method, a statement we believe to be true must either be proved or else added to our list of axioms. And for reasons of logical economy and elegance, we wish to rely on as few axioms as possible.

In our treatment of the natural numbers we accepted the well-ordering principle as an axiom about the natural numbers. But, in addition, we accepted as given facts your understanding of elementary arithmetic: that is, addition, subtraction, and multiplication. This won't lead to any difficulties later, but let's see how we can develop $\mathbb{N}$ if we wanted to be very rigorous.

The following axioms are called the *Peano axioms* for the natural numbers.

Let $\mathbb{N}$ be a set with a special element $0 \in \mathbb{N}$ and a function $S : \mathbb{N} \to \mathbb{N}$ that satisfy the following:

Axiom 1. For every natural number x, $S(x) \neq 1$.

Axiom 2. For every natural numbers x and y, if $S(x) = S(y)$, then $x = y$.

Axiom 3. (Axiom of Induction) If T is a set of natural numbers with the properties

A: 1 belongs to T

B: If x belongs to T, then so does $S(x)$

then T contains all the natural numbers.

The function S is called the *successor* function. Note that in the Peano axioms, we assume that induction holds for the natural numbers (Axiom 3).

The way to proceed in this system would be first to establish a series of theorems, most of which you would agree are obvious, but we'd want to be absolutely sure of their truth. The first few theorems would be as follows:

Theorem 1: If $x \neq y$, then $S(x) \neq S(y)$.

Theorem 2: $S(x) \neq x$.

Theorem 3: If $x \neq 1$, then there exists one u such that $x = S(u)$.

After these three theorems, we can now define *inductively* the addition of two natural numbers:

To every pair of natural numbers x and y, we assign in a natural number, called $x + y$, such that

A: $x + 1 = S(x)$, for every x,

B: $x + S(y) = S(x + y)$, for every x and y.

We quite naturally call $x + y$ the sum of x and y.

From this definition, we could then establish that addition is associative and commutative, and various other facts about natural numbers that are well known to us.

Exercises

1. Describe how to do a tiling of the 2^n-by-2^n squares with

one box removed using L-shaped tiles. Your method should be inductive in nature.

2. Prove using mathematical induction that for all positive integers n,

$$1 + 2 + 3 + \cdots + n = \frac{n(n+1)}{2}.$$

3. Prove using mathematical induction that for all positive integers n,

$$1^2 + 2^2 + 3^2 + \cdots + n^2 = \frac{n(2n+1)(n+1)}{6}.$$

4. Prove using mathematical induction that for all positive integers n,

$$1 \cdot 1! + 2 \cdot 2! + \cdots + n \cdot n! = (n+1)! - 1.$$

5. Prove using mathematical induction that for all positive integers n,

$$2^1 \cdot 1 + 2^2 \cdot 2 + 2^3 \cdot 3 + \cdots + 2^n \cdot n = 2 + (n-1)2^{n+1}.$$

6. Prove that $5 + 8 + 11 + ... + (3n+2) = \frac{1}{2(3n^2+7n)}$.
7. A football quarterback's contract states that for the first game the team wins, the quarterback will get a \$1000 bonus, for the second game won, a \$2000 bonus, for the third win a \$3000 bonus, and so on. If the team wins 10 games, how much will the total bonus be?
8. Prove $n! > 3^n$ for all $n \geq 7$.
9. You probably recall from your previous mathematical work the *triangle inequality*: For any real numbers x and y,

$$|x + y| \leq |x| + |y|.$$

Accept this as given (or see a calculus text to recall how it is proved). Generalize the triangle inequality, by proving that

$$|x_1 + x_2 + \cdots + x_n| \leq |x_1| + |x_2| + \cdots + |x_n|,$$

for any positive integer n. *Hint:* Use mathematical induction and the triangle inequality.

10. Prove that $2^n < n!$ for all $n \geq 4$. (Note that here your base case for induction is when $n = 4$. You can easily check that this statement is false for $n = 1$, 2, and 3.)

11. Prove that for all positive integers n,

$$1^3 + 2^3 + \cdots + n^3 = \left(\frac{n(n+1)}{2}\right)^2.$$

12. Prove the familiar geometric progression formula:

$$a + ar + ar^2 + \cdots + ar^{n-1} = \frac{a - ar^n}{1 - r},$$

where $r \neq 1$.

13. Prove that for all positive integers n,

$$\frac{1}{1 \cdot 2} + \frac{1}{2 \cdot 3} + \cdots + \frac{1}{n(n+1)} = \frac{n}{n+1}.$$

14. By trial and error, try to find the smallest positive integer expressible as $12x + 28y$, where x and y are allowed to be any integers.

15. Consider the sequence $\{a_n\}$ defined inductively as follows:

$$a_1 = a_2 = 1, \quad a_{n+2} = 2a_{n+1} - a_n.$$

Use mathematical induction to prove that $a_n = 1$ for all natural numbers n.

16. If $a_1 = 5$ and $a_n = 2a_{n-1}$, for $n > 1$, then prove that $a_n = 5 \cdot 2^{n-1}$ for all $n \geq 1$.

17. If $t_1 = 1$ and $t_n = 2t_{n-1} + 1$, for $n > 1$, then prove that $t_n = 2^n - 1$ for all $n \geq 1$.

18. If $a_1 = 2$ and $a_n = 3a_{n-1} + 1$, for $n > 1$, then prove that $a_n = (5 \cdot 3^{n-1} - 1)/2$, for $n \geq 1$.

19. If $b_1 = 24$ and $b_n = 3b_{n-1} + 1$, for $n > 1$, then prove that $b_n = 3^{n-1}$ for all $n \geq 1$.

20. Consider the sequence $\{a_n\}$ defined inductively as follows:

$$a_1 = 5, \quad a_2 = 7, \quad a_{n+2} = 3a_{n+1} - 2a_n.$$

Use mathematical induction to prove that $a_n = 3 + 2^n$ for all natural numbers n.

* 21. Recall that $\binom{n}{k} = n!/(k!(n-k)!)$.

a. Prove that

$$\binom{n}{k} = \binom{n-1}{k} + \binom{n-1}{k-1},$$

where $k < n$ and

$$\binom{n}{k} = \frac{n!}{(n-k)!k!}.$$

Hint: You do not need induction to prove this. Bear in mind that $0! = 1$.

b. Use part (a) and induction to prove the binomial theorem: For nonnegative n,

$$(x + y)^n = \sum_{k=0}^{n} \binom{n}{k} x^{n-k} y^k.$$

22. Prove that $\frac{(2n)!}{2^n}$ is an integer for all natural numbers n.
23. Suppose $n > 2$. How many diagonals in a regular polygon with n sides are there? Prove your answer using induction.
24. Prove that all integers greater than or equal to 12 can be written as a sum of 5's and 3's.
25. Prove that all numbers greater than or equal to 20 can be written as a sum of 5's and 6's.
26. Suppose n is an odd number. Consider the statement "all numbers greater than or equal to N can be written as a sum of 2's and n's." What is the smallest value for N (written in terms of n) for which this statement is true?
27. At McDonald's restaurants, you can purchase Chicken McNuggets in boxes of 6, 9, and 20. What is the smallest N for which you can buy exactly n McNuggets for all $n \geq N$? Prove your assertion. (*Hint*: N is greater than 30.)
28. Suppose there is a country with coins of denominations 1, 3, 8, and 10 cents. Suppose a store clerk runs out of 1 cent coins but has an unlimited supply of the other three coins. Show that the clerk can still make change for all values greater than or equal to 8 cents.
29. Suppose the same clerk in Exercise 28 also runs out of 10 cent coins, leaving only 3 and 8 cent coins. Show that the clerk can make change for any value 14 cents or higher.
30. Again, suppose the clerk has only 3 and 10 cent coins. Show that the clerk can make change for any value 19 cents or higher.
31. Consider the statement "Every number greater $\geq N$ can be written as a sum of a's and b's." For which of the following values of a and b is this statement true? If false, explain why. If true, find the smallest such N and prove it is so.

a. 2 and 5
b. 4 and 6
c. 2 and 4
d. 3 and 7

* 32. There are n teams that play a round-robin tournament where there are no ties. (That is, every team plays every other team.) A ranking is an arrangement of the teams in a list, $t_1, t_2, \ldots, t_n$, where t_1 beats t_2, t_2 beats t_3, and so on. Prove that a ranking of teams is always possible regardless of the outcome of the tournament. (Of course, it may be true that for some outcomes, there may be more than one ranking.)

33. Criticize the following "proof" showing that all cows are the same color.
It suffices to show that any herd of n cows has the same color. If the herd has but one cow, then trivially all the cows in the herd have the same color. Now suppose that we have a herd of n cows and $n > 1$. Pick out a cow and remove it from the herd, leaving $n - 1$ cows; by the induction hypothesis these cows all have the same color. Now put the cow back and remove another cow. (We can do so because $n > 1$.) The remaining $n - 1$ again must all be the same color. Hence, the first cow selected and the second cow selected have the same color as those not selected, and so the entire herd of n cows has the same color.

* 34. Prove that the principle of mathematical induction implies the well-ordering principle. (This shows that these two principles are logically equivalent, and so from an axiomatic point of view it doesn't matter which we assume is an axiom for the natural numbers.) *Hint for the proof:* Suppose that S is a subset of $\mathbb{N}$ that does not have a least element. Prove that S is empty, by using induction on the set $\mathbb{N} \setminus S$.

* 35. Prove that addition is associative and commutative.

36. Give an inductive definition for multiplying two natural numbers.

* 37. Show that multiplication, as defined in the last exercise, is associative and commutative.

38. Give an inductive definition for n^m.

Programming Exercises

1. Write a program that sorts a set of integers and keeps track of the number of comparisons.
2. Write a program that displays a table with three columns: n, n^2, and the sum of the squares of all natural numbers up to n.

Chapter 5
Number Theory

Number theory was once thought to be the "purest of the pure" mathematics. But with the advent of the computer and sophisticated encryption techniques, number theory has become known as an applied area of mathematics. Certainly a basic understanding of the principles of number theory—prime factorization and modular arithmetic—should be part of every computer scientist's background. Here we will study enough number theory to understand an encryption method called RSA encryption (named after its inventors Rivest, Shamir, and Adleman). This method relies on some old and well-known theorems of number theory once only of interest to mathematicians. As always, the mathematics we'll learn along the way has applications to situations other than the one we've chosen here as an example.

If we take the additive inverses of the natural numbers, $\mathbb{N}$, together with $\mathbb{N}$ and 0, we get the set of the *integers*: $\ldots, -2, -1, 0, 1, 2, 3, \ldots$. Note that if we add or multiply two natural numbers, we get another natural number; we say the set of natural numbers is closed under addition and multiplication. However, $\mathbb{N}$ is not closed under subtraction or division. (The integers are closed under addition, multiplication, and subtraction but not division. Some values of $\mathbb{Z}$ do divide into others, however.)

If $a, b \in \mathbb{N}$, we say that a *divides* b if there is a $k \in \mathbb{N}$ so that $a \cdot k = b$. We write $a|b$ in this case. So $3|12$ since $3 \cdot 4 = 12$. If no such natural number k exists, we say a *does not divide* b and write $a \not| b$. For example, $3 \not| 10$. If $a|b$, we also say a is a *factor* of b or b is a *multiple* of a. Sometimes, to simplify the language, if we refer to a as a number, we mean a is a natural number.

✓ List all natural numbers less than 10 that divide 10. Does $10|10$?

If $a \neq 0$, then $a|0$ since $0 = 0 \cdot a$. So we'll expand our definition of "divides" to include this case. Thus our revised definition of divides is: If $a \in \mathbb{N}$ and $b \in \mathbb{N} \cup \{0\}$, we say that *a divides* b if there is a $k \in \mathbb{N} \cup \{0\}$ so that $a \cdot k = b$.

The Division Theorem

A bedrock theorem in number theory is the Division Theorem. This result is one you have known in some form since grade school, but that doesn't diminish its importance. You learn early on that, although not every number divides every other number, we can perform division by permitting a remainder. For example, when we divide 320 by 12 we get a quotient of 26 and a remainder of 8. We can write this succinctly as $320 = 12 \cdot 26 + 8$. We formally describe this situation in the following theorem:

The Division Theorem. *If a and b are two natural numbers, then there exist unique natural numbers q and r with $0 \leq r < b$, such that $a = b \cdot q + r$.*

We will not prove this theorem here. Note the key points of the theorem: The number r is between 0 and b, inclusive of 0, and the two numbers q and r are *unique.* That is, there are no other numbers that have the properties given for q and r.

For example, if $a = 39$ and $b = 4$, then $q = 9$ and $r = 3$, since $39 = 4 \cdot 9 + 3$. If $a = 5$ and $b = 13$, then $q = 0$ and $r = 5$, since $5 = 13 \cdot 0 + 5$. We use the letter q to indicate *quotient* and r to indicate *remainder.*

Greatest Common Divisors

While it is a time-consuming problem in general to factor a number, especially a large one, it turns out to be easy to determine a common factor of two numbers. In fact we can easily and efficiently find the largest common divisor of two numbers using an algorithm dating from the fourth century *B.C.* due to Euclid.

Given two natural numbers, a and b, a *common divisor* of a and b is any number d that divides both a and b. Notice that 1 is a common divisor of any pair of numbers. Two integers may have many common divisors. For instance, 12 and 18 have common divisors of $1, 2, 3, 4$, and 6. If d is the largest common divisor of a and b, we call d the *greatest common divisor of a and b* (or simply the *gcd of a and b*) and write $d = \gcd(a, b)$. For example, $\gcd(42, 18) = 6$ while $\gcd(24, 15) = 3$. If $\gcd(a, b) = 1$, we say that a and b are *relatively prime.* (Two integers being relatively prime does not imply that either one is prime. For example, $\gcd(4, 15) = 1$.)

✓	What is the value of $\gcd(20, 45)$? Of $\gcd(21, 16)$?

Consider the special case, of $\gcd(a, 0)$, if $a \neq 0$. Since every nonzero number divides 0 (since $0 = 0 \cdot a$), $\gcd(a, 0) = a$.

Why should any pair of numbers have a gcd? We've already noticed that 1 is a common divisor of any pair a and b, so the set of all common divisors of a and b is not empty. Now the largest possible value for $\gcd(a, b)$ is the smaller of a and b. Thus, the set of all common divisors of a and b is a finite set and so there must be a greatest element in it.

To find $\gcd(a, b)$, we could simply start at the smaller of a and b and work our way down to 1, stopping at the first number we find that is indeed a common divisor of a and b. But this could be a lot of work if a and b are large and relatively prime or if their gcd is small. Fortunately, there is a better method: Euclid's algorithm. (This first appears in Euclid's *Elements* as Proposition 2 of Book 7.)

Euclid's algorithm is probably best presented by example. Notice in the example that the algorithm repeatedly finds the quotient and remainder of a succession of numbers and writes out the corresponding equations using the Division Theorem.

Let's find gcd(285, 255):

$$285 = 255 \cdot 1 + 30$$
$$255 = 30 \cdot 8 + 15$$
$$30 = 15 \cdot 2 + 0$$

We conclude that $\gcd(285, 255) = 15$.

To compute $\gcd(a, b)$, this algorithm first writes $a = b \cdot q + r$ as guaranteed by the Division Theorem (so $0 \leq r < b$). Then repeat this step setting $a = b$ and $b = r$. We repeat this process until $r = 0$. The last value for b we obtained is the gcd of our original a and b. (Note that we don't need to compute q in each step.) The algorithm is as follows:

```
function gcd(a,b)
% require:  a & b positive integers
% ensure:  returns gcd of a & b
repeat
  r = a mod b;
  a = b;
  b = r;
until (r = 0);
return(a);
end function;
```

Let's do a more lengthy example by finding gcd(110, 42):

$$110 = 42 \cdot 2 + 26$$
$$42 = 26 \cdot 1 + 16$$
$$26 = 16 \cdot 1 + 10$$
$$16 = 10 \cdot 1 + 6$$
$$10 = 6 \cdot 1 + 4$$
$$6 = 4 \cdot 1 + 2$$
$$4 = 2 \cdot 2 + 0$$

So $\gcd(110, 42) = 2$. Why does this algorithm work? It follows from the fact that if $a = b \cdot q + r$, then $\gcd(a, b) = \gcd(b, r)$. To show this, we need only check that *every* common divisor of a and b is also a common divisor of b and r, and vice versa, for then the greatest element of this set will be the gcd we want. But if $d|a$ and $d|b$, then $d|r$ since we can write $r = a - b \cdot q$. Likewise, if $d|b$ and $d|r$, then $d|a$ since $a = b \cdot q + r$.

Now to see why this fact shows that Euclid's algorithm gives the desired gcd, let's look at the last example we did. By our first line of computation (just the Division Theorem) we see that $\gcd(110, 42) = \gcd(42, 26)$. But the next line shows that $\gcd(42, 26) = \gcd(26, 16)$, so $\gcd(110, 42) = \gcd(26, 16)$. Following this through, we see that $\gcd(110, 42) = \gcd(4, 2)$. This last value is obviously 2 since $2|4$, as we see because the last remainder is 0.

> ✓ Use Euclid's algorithm to compute $\gcd(12, 39)$ and $\gcd(45, 12)$.

There is one last thing that a slight modification of Euclid's algorithm allows us to compute. If $\gcd(a, b) = d$, then we can find integers m and n so that $d = a \cdot m + b \cdot n$. This is called the *gcd identity.* For example, $\gcd(110, 42) = 2$ and we can write $2 = 110 \cdot -8 + 42 \cdot 21$. In this case we say we can *write 2 as a linear combination of 110 and 42.* While it is not obvious how to compute m and n, Euclid's algorithm allows us to find these two integers m and n. The usefulness of this is not apparent now, but will be when we get to RSA encryption. We will illustrate using the last example of Euclid's algorithm given previously. A recursive version of the algorithm will be given in the recursion chapter.

The scheme for finding the values for a and b is quite simple. We start at the next to last line of the computation in Euclid's algorithm, $6 = 4 \cdot 1 + 2$. Since $2 = \gcd(6, 4) = \gcd(110, 42)$, solving for 2 gives us

$$2 = 6 \cdot 1 + 4 \cdot -1.$$

Now we can solve the previous line for 4, substitute it into this last equation, and, with a little algebra, we have 2 as a linear combination of 10 and 6:

$$\begin{aligned} 2 &= 6 \cdot 1 + (10 + 6 \cdot -1) \cdot -1 \\ &= 10 \cdot (-1) + 6 \cdot 2. \end{aligned}$$

Now we can solve the next line up for 6, substitute and write 2 as a linear combination of 16 and 10. We continue working our way up until we write 2 as a linear combination of 110 and 42. You should do this to see that you get the same numbers we did.

Primes

It is easy to see that 1 divides every number (since $1 \cdot n = n$ for every number n) and each number divides itself (since $n \cdot 1 = n$ for every number n). A number p is called *prime* if $p \neq 1$ and the only numbers that divide p are 1 and p itself. In other words, primes have the minimal number of divisors possible. Note that 1 is not prime. If a number (greater than 1) is not prime, we call it *composite*.

> ✓ The first ten primes are 2, 3, 5, 7, 11, 13, 17, 19, 23, and 29. What are the next five primes?

An important property of primes is given in the following theorem.

Theorem. *If p is prime and $p|ab$, then either $p|a$ or $p|b$.*

Proof: If $p|a$ then we're done. Suppose $p \nmid a$. We need to show $p|b$. Let d be a common divisor of p and a. Since p is prime, d is either p or 1. But $p \nmid a$, so $d = 1$. The gcd identity says there exist integers x and y such that $1 = ax + py$. Multiplying by b, we have $b = abx + pby$. Since $p|ab$ and $p|pby$, $p|(abx + pby)$. That is, $p|b$ as required. //

For example, $5|80$ and since $80 = 20 \cdot 4$ and 5 is prime, then either $5|20$ or $5|4$ (or both). Obviously, $5|20$. Generalizing this theorem, we note the following:

Theorem. *If p is prime and $p|a_1a_2\cdots a_n$ then $p|a_i$ for some i.*

Proof: This is a straightforward induction on n and is left as an exercise. //

Primes are the basic building blocks of the natural numbers in the following sense: Every natural number (> 1) is either prime or can be written *uniquely* as the product of primes. By *uniquely* we mean that there is only one collection of primes whose product is the number in question. (We ignore the order in which we list the primes, since multiplication is commutative.) This fact is expressed in the following theorem:

Fundamental Theorem of Arithmetic. *Every natural number greater than 1 is either prime or can be written uniquely as a product of prime numbers.*

Proof: We prove this theorem in two parts: First we show that each number greater than 1 can be written as a product of primes. Then we show that when a number is written as a product of primes, this product is unique (up to the order this product is written). We prove the first part by induction.

Let $X = \{n : n \text{ is prime or can be factored into primes}\}$. It is clear that 2 is prime and so satisfies this theorem. Thus, $2 \in X$. Now let $n > 2$ and suppose that all numbers less than n and greater than 1 satisfies this theorem. (That is, all these numbers are in X.) Now we need to show that n satisfies this theorem.

If n is prime, we are done. Otherwise, n is composite and so we can write $n = a \cdot b$, where a and b are two numbers greater than 1 and less than n. Now by our induction hypothesis, both a and b are in X; that is, they are either prime or can be written as a product of primes. But then it follows that $n = a \cdot b$ can be written as a product of primes; it is simply the product of all the primes that make up the factorization of a with all the primes that make up the factorization of b. Thus $n \in X$ and so, by induction, $X = \{2, 3, \ldots\}$, as desired.

It remains to show that this product is unique. That is, there is no other collection of primes whose product is n. We again use induction, but our argument will be a little less formal than in previous cases. Specifically, we won't formally identify the set X. It's still in the background but is understood. This way of writing an induction proof is common.

Assume that $n = a_1a_2 \cdots a_m = b_1b_2 \cdots b_k$ are two factorizations of n into primes. We wish to show that $m = k$ and that the b_j may be rearranged so that $a_i = b_i$ for $i = 1, 2, \ldots, m$. We now use induction on m, the number of prime factors in the first factorization. If $m = 1$, then n is prime and so $k = 1$ also. So we assume $m > 1$. By the previous theorem, since a_1 is prime and divides n, then a_1 divides one of the b_j. By renumbering if necessary, we can assume that a_1 divides b_1. But b_1 is also prime so $a_1 = b_1$. So dividing both sides by a_1 ($=b_1$) we get that $a_2a_3 \cdots a_m = b_2b_3 \cdots b_k$. But then by our induction hypothesis $m - 1 = k - 1$ and by renumbering the b_j as necessary, $a_i = b_i$ for $i = 2, 3, \ldots, m$. This proves the theorem. //

The Fundamental Theorem of Arithmetic is an existence theorem. That is, it establishes that every natural number (greater than one) can be factored into primes but does not tell you how to go about finding the factorization. Indeed, factoring is a very difficult problem in general and is the focus of much current research. The security of the RSA encryption method depends on factoring being a time-consuming problem. A breakthrough fast factoring method would make this encryption method suspect, at least for the size of keys currently being used. Paradoxically, determining whether or not a number is prime is an easier problem than factoring. That is, you may determine fairly easily that a number is composite but not be able to factor it! Later, we will see a theorem that does exactly that, without giving a clue as to how to factor the number.

The following algorithm is a straightforward method to

determine whether or not a number n (> 1) is prime. We need only check possible divisors up to $\lfloor\sqrt{n}\rfloor$. (Why this is so is left as an exercise.)

```
function Prime(n)
% require:  n is an integer > 1
% ensure:  function returns true
%          if and only if n is prime
IsPrime = true;
for a = 2 to ⌊√n⌋ do
  if (a divides n) then
    IsPrime = false;
  endif;
endfor;
return(IsPrime);
end function;
```

> ✓ Use the function `Prime` to determine if 101 is prime and if 91 is prime.

We could save some computation if we broke out of the `for` loop once we found that a divides n; some programming languages allow us to do that easily. You would use the `mod` function in your language to determine if a divides n. For instance, in Pascal, the boolean statement "a divides n" would be written "`(n mod a = 0)`" and in C and C++, % is the mod operater so "a divides n" would be written "`(n % a == 0)`". A further refinement would be to only check those numbers a that are prime, since n is composite if and only if it has a prime factor. You might ponder the difficulties of doing this without making the algorithm run slower.

For relatively small numbers, we can quickly determine all the primes no greater than n using the *sieve of Erastothanes*, which is one of the oldest recorded algorithms. An alternative method is to call the preceding algorithm for each number, which repeats many calculations unnecessarily, something the sieve of Erastothanes avoids. This algorithm implements the sieve by using a boolean array IsPrime[2..n]. At the completion of the algorithm, IsPrime[i] is true if i is prime and false otherwise.

```
function Sieve(n, IsPrime[])
% require:  n is an integer > 1,
%           IsPrime[2..n] is a boolean array
% ensure:   IsPrime[i] = true if i is prime and
%          IsPrime[i] = false if i is not prime
%          for all i s.t.  1 < i < n+1
   % initially mark all numbers as "prime"
for i = 2 to n do
  IsPrime[i] = true;
endfor;
for i = 2 to ⌊√n⌋ do
  % if i is prime
  if (IsPrime[i]) then
    % mark all multiples of i as "composite"
    a = 2 * i;
    while (a ≤ n) do
      IsPrime[a] = false;
      a = a + i;
    endwhile;
  endif;
endfor;
end function;
```

As an example, let's use `Sieve` to find all primes less than 38. Since $\lfloor\sqrt{38}\rfloor = 6$, the main loop in the algorithm repeats for `i` from 2 through 6. We'll list all the elements from 2 through 37 and mark those n where we've set `IsPrime[`n`]=false`. We'll mark by putting a line over the number, such as $\overline{20}$. We show the list after each pass.

after `i = 2`:

$$2\ 3\ \overline{4}\ 5\ \overline{6}\ 7\ \overline{8}\ 9\ \overline{10}\ 11\ \overline{12}\ 13\ \overline{14}\ 15\ \overline{16}\ 17\ \overline{18}\ 19\ \overline{20}$$

$$21\ \overline{22}\ 23\ \overline{24}\ 25\ \overline{26}\ 27\ \overline{28}\ 29\ \overline{30}\ 31\ \overline{32}\ 33\ \overline{34}\ 35\ \overline{36}\ 37$$

after `i = 3`:

$$2\ 3\ \overline{4}\ 5\ \overline{6}\ 7\ \overline{8}\ \overline{9}\ \overline{10}\ 11\ \overline{12}\ 13\ \overline{14}\ \overline{15}\ \overline{16}\ 17\ \overline{18}\ 19\ \overline{20}$$
$$\overline{21}\ \overline{22}\ 23\ \overline{24}\ 25\ \overline{26}\ \overline{27}\ \overline{28}\ 29\ \overline{30}\ 31\ \overline{32}\ \overline{33}\ \overline{34}\ 35\ \overline{36}\ 37$$

after `i = 4`: (no change since `IsPrime[4] = false`)

$$2\ 3\ \overline{4}\ 5\ \overline{6}\ 7\ \overline{8}\ \overline{9}\ \overline{10}\ 11\ \overline{12}\ 13\ \overline{14}\ \overline{15}\ \overline{16}\ 17\ \overline{18}\ 19\ \overline{20}$$
$$\overline{21}\ \overline{22}\ 23\ \overline{24}\ 25\ \overline{26}\ \overline{27}\ \overline{28}\ 29\ \overline{30}\ 31\ \overline{32}\ \overline{33}\ \overline{34}\ 35\ \overline{36}\ 37$$

after `i = 5`:

$$2\ 3\ \overline{4}\ 5\ \overline{6}\ 7\ \overline{8}\ \overline{9}\ \overline{10}\ 11\ \overline{12}\ 13\ \overline{14}\ \overline{15}\ \overline{16}\ 17\ \overline{18}\ 19\ \overline{20}$$
$$\overline{21}\ \overline{22}\ 23\ \overline{24}\ \overline{25}\ \overline{26}\ \overline{27}\ \overline{28}\ 29\ \overline{30}\ 31\ \overline{32}\ \overline{33}\ \overline{34}\ \overline{35}\ \overline{36}\ 37$$

after `i = 6`: (no change since `IsPrime[6] = false`)

$$2\ 3\ \overline{4}\ 5\ \overline{6}\ 7\ \overline{8}\ \overline{9}\ \overline{10}\ 11\ \overline{12}\ 13\ \overline{14}\ \overline{15}\ \overline{16}\ 17\ \overline{18}\ 19\ \overline{20}$$
$$\overline{21}\ \overline{22}\ 23\ \overline{24}\ \overline{25}\ \overline{26}\ \overline{27}\ \overline{28}\ 29\ \overline{30}\ 31\ \overline{32}\ \overline{33}\ \overline{34}\ \overline{35}\ \overline{36}\ 37$$

So we see that the primes less than 38 are 2, 3, 5, 7, 11, 13, 17, 19, 23, 29, 31, and 37.

✓ Use the sieve to find all two-digit primes.

Of course, neither of these algorithms is effective if n is large. The first algorithm for determining if n is prime takes $\sqrt{n}$ steps. This works effectively for small numbers that can be stored in a single word on a computer, but if n had 100 digits or so, $\sqrt{n}$ would have about 50 digits, still a very large number, making the first algorithm intractable. Other methods must be used to determine if these very large numbers are prime.

Modular Arithmetic

Given a fixed number n (which we'll call the *base*), there are many numbers that have the same remainder when divided by n. For example, $1, 6, 11, 16$, and 21 all have a remainder of 1 when divided by 5. We say all these numbers are equal to 1 "modulo 5" (or just "mod 5"). To say that these numbers are equal mod 5 means they are all equivalent under the relation $\sim_5$

defined by $a \sim_5 b$ iff $5|(a-b)$. (You can easily show that this is an equivalence relation.) Thus all the integers equal to 1 mod 5 are in the same equivalence class mod 5. The remainders mod 5 can be 0, 1, 2, 3, or 4. These are called the *residues modulo 5*. Hence there are five equivalence classes mod 5, each with its unique residue. The residues are the smallest nonnegative representatives in each of the classes.

This can be generalized. If n is any integer greater than 1, and the relation $\sim_n$ is defined by $a \sim_n b$ iff $n|(a-b)$, then $\sim_n$ is an equivalence relation. Furthermore, this relation has n equivalence classes with residues $0, 1, 2, \cdots, n-1$.

We can do arithmetic using these residues in a way similar to the usual arithmetic with integers. Indeed, the five residues of 5 are sometimes referred to as the "integers mod 5" and are denoted $\mathbb{Z}_5$. Here's how you add in $\mathbb{Z}_5$: To add two residues, simply take two integers whose residue mod 5 are the residues you wish to add, add these integers, and compute the residue of the resulting sum. For instance, to compute $3+4 \pmod 5$ we take any integer whose residue is 3 (say 8) and any integer whose residue is 4 (say 4), add them (we get 12), and compute the residue of this sum mod 5 (we get 2). So $3+4=2 \pmod 5$. You might wonder what happens if we make another choice of integers whose residues are 3 and 4; say we take 13 and 29. The sum is 42, which is $2 \pmod 5$. Indeed, regardless of our choice of integers with residues 3 and 4, the residue of the sum (mod 5) will always be 2. Thus our definition of addition is well-defined. In practice, you usually pick the smallest integers possible (that is, the residues) when doing this calculation. Keep in mind that we are really doing arithmetic on the equivalence classes, but doing so using representatives of those classes.

Subtraction is handled a little differently. In the integers, we can think of subtraction as adding the additive inverse of a number. That is, we can think of $m+n$ as $m+(-n)$, where $-n$ is the number that we add to n to get 0. (We call $-n$ the *additive inverse* of n.) The number 0 is the *additive identity* since $a + 0 = a \pmod b$ for all residues a. When we think of subtraction this way, we see that to carry it over to Z_b we need to find the value of $-a$ in $\mathbb{Z}_b$ for every $a \in \mathbb{Z}_b$, where $-a \pmod b$ is the residue we add to a to get $0 \pmod b$. For example, in $\mathbb{Z}_5$, $-3 = 2$ since $3 + 2 = 0$. Knowing this, we can now subtract 3 from the other residues: $0 - 3 = 2$ (since $0-3 = 0+2 = 2$), $1-3 = 3$ (since $1-3 = 1+2 = 3$), $2-3 = 4$ (since $2-3 = 2+2 = 4$), $3-3 = 0$ (since $3-3 = 3+2 = 0$), and $4-3 = 1$ (since $4-3 = 4+2 = 1$). Similarly, we can compute the additive inverses of the other residues: $-0 = 0$, $-1 = 4$, $-2 = 3$ and $-4 = 1$. We can now do subtraction mod 5. For example, $4-2 = 4+3 = 2 \pmod 5$ and $2-4 = 2+1 = 3 \pmod 5$. When doing arithmetic in a given modulus, we usually tack on $\pmod b$ after the computation to make clear in what modulus we are doing the arithmetic.

In general, it is very easy to compute $-a \pmod b$; the value is simply $b-a$ (assuming that $0 \leq a < b$). To verify this, recall that $-a$ is what we add to a to get 0 (where all arithmetic is in $\mathbb{Z}_b$, of course). But, $a + (b-a) = b = 0 \pmod b$. And so, $-a = b-a$. Thus, $-23 \pmod{26} = 3$ and $-15 \pmod{100} = 85$.

This allows us to solve certain equations in $\mathbb{Z}_b$. Consider the following: $5 + x = 2 \pmod 7$. We wish to solve for x. The method is the same as usual arithmetic in the integers; we simply add -5 to both sides. But in $\mathbb{Z}_7$, $-5 = 2$. Thus, $2+5+x = 2+2 \pmod 7$, or $x = 4 \pmod 7$. We will examine multiplication and division later.

A Cryptological Example

Cryptology is the study of taking a message, changing it into a coded message, transmitting that coded message, and finally uncoding it to reveal the original message. The simple substitution encryptions you see in newspaper cryptograms are examples of this. Changing the original message (called the plaintext) into the coded message (called the ciphertext) is call *encryption*. The inverse operation, recovering the plaintext from the ciphertext, is call *decryption.* If you know the method of encryption, the extra information you need in order to encrypt the plaintext is called the *encryption key* and the extra information you need in order to decrypt the ciphertext is called the *decryption key.* RSA encryption is the most popular of the methods of encryption called public-key encryption. This means that the encryption key is made public, but even this does not allow someone to figure out the decryption key! The fact that such methods even exist are surprising.

Before learning about public-key encryption, we will look at a simpler method. A very simple encryption method is known as a shift cipher. Suppose the plaintext is `MEET ME AT NOON`. We will assume that we only use the 26 letters of the alphabet and do not distinguish between uppercase and lowercase letters. The first step is to translate each letter into a number. Since we have 26 letters, it is natural to do our arithmetic in $\mathbb{Z}_{26}$. So let $\mathtt{A} = 0$, $\mathtt{B} = 1, \ldots$, $\mathtt{Z} = 25$. Then our plaintext becomes the string of numbers 12 4 4 19 12 4 0 19 13 14 14 13. We consider these numbers as residues mod 26 and encrypt one number at a time. Using the shift method of encryption, we must pick a shift—this is the *encryption key* for this method. Let's choose 15 as our shift. If we let p be our plaintext letter and c be our corresponding ciphertext letter, then we compute c from p with the formula $p + 15 = c \pmod{26}$. Thus our ciphertext is 1 19 19 8 1 19 15 8 2 3 3 2, which would be converted to the letters `BTTIB TPICD DC`. It is common practice to group the ciphertext letters in fives.

We could use any of the 26 shifts to encrypt this message. (However, a shift of 0, which is equivalent to a shift of 26, wouldn't be very secure since the ciphertext would be the same as the plaintext!) Historically, a shift of 3 is known as the Caesar Cipher after Julius Caesar, who used it.

> ✓ Encrypt the message `THIS MESSAGE IS ENCRYPTED` using the aforementioned scheme with a shift of 5.

To decrypt, the receiver of the message needs to recover the plaintext from the ciphertext. The receiver must know the encryption key (15 here) and from that can compute the decryption key, which is another shift. The receiver can do this by solving the encryption equation, $p + 15 = c \pmod{26}$, for p, noting that $-15 = 11 \pmod{26}$: $p + 15 + 11 = c + 11 \pmod{26}$, or $p = c + 11 \pmod{26}$. Thus, the decryption key is 11. Using this formula, the receiver can now recover the original plaintext, as you can check.

> ✓ What is the decryption key for the message you encrypted in the preceding example? Use it to decrypt the ciphertext you created. Of course, you should recover the original plaintext.

Of course, this is not a very secure encryption method since there are only 26 possible shifts when using mod 26 (really only 25 since a shift of 0 doesn't conceal much!). Someone intercepting the message could easily try all 25 shifts and find the plaintext. With the aid of a computer, decryption could be done very quickly.

Modular Multiplication and Division

Multiplication mod b is similar to addition: $m \cdot n \pmod{b}$ is the residue mod b of $m \cdot n$. Thus $5 \cdot 7 = 8 \pmod{9}$ and $5 \cdot 7 = 5 \pmod{10}$. Division is done in a manner similar to subtraction. As 0 is the additive identity of $\mathbb{Z}_b$, 1 is the *multiplicative identity* of $\mathbb{Z}_b$ since $1 \cdot a = a \pmod{b}$ for all a in $\mathbb{Z}_b$. When doing arithmetic with real numbers, $m/n = m \cdot (1/n)$, where $1/n$ is the *multiplicative inverse* of n; that is, $1/n$ is the number we multiply n by to get 1. We commonly write n^{-1} for $1/n$. We carry this idea over to $\mathbb{Z}_b$. Let's consider

$\mathbb{Z}_5$: $2^{-1} = 3 \pmod 5$ since $2 \cdot 3 = 1 \pmod 5$. Similarly, $1^{-1} = 1 \pmod 5$, $3^{-1} = 2 \pmod 5$ and $4^{-1} = 4 \pmod 5$, since $1 \cdot 1 \pmod 5$, $3 \cdot 2 = 1 \pmod 5$ and $4 \cdot 4 = 1 \pmod 5$. As in the usual arithmetic, 0 has no multiplicative inverse, since 0 times every number is 0.

This allows us to solve more complicated equations in $\mathbb{Z}_5$ in a manner similar to the way we solve them in our usual arithmetic. For example, let's find x that satisfies $3x + 2 = 4 \pmod 5$. Of course, we know that $-2 = 3 \pmod 5$ and $3^{-1} = 2 \pmod 5$, so

$$\begin{gathered} 3x + 2 = 4 \pmod 5 \\ 3x + 2 + 3 = 4 + 3 \pmod 5 \\ 3x = 2 \pmod 5 \\ 2 \cdot 3x = 2 \cdot 2 \pmod 5 \\ x = 4 \pmod 5. \end{gathered}$$

Thus, $x = 4$ is the (unique) solution to our equation. You can easily check that this value is correct.

If we try to compute the multiplicative inverses of the nonzero elements of $\mathbb{Z}_6$, things are very different. We find that $1^{-1} = 1 \pmod 6$, as usual, and $5^{-1} = 5 \pmod 6$, but no multiplicative inverse exists for 2, 3, or 4. (Try the possibilities and see for yourself.) In turns out that a multiplicative inverse exists for $a \pmod b$ if and only if a and b are relatively prime. (We will not prove this fact here, but you may want to think about how to do so.) In $\mathbb{Z}_{10}$, 1, 3, 7, and 9 have multiplicative inverses but 2, 4, 5, 6, and 8 (as well as 0) do not. In particular, if p is prime, then all nonzero elements of $\mathbb{Z}_p$ have multiplicative inverses.

> ✓ Find $5^{-1} \pmod 7$ and $3^{-1} \pmod 8$. Which numbers in $\mathbb{Z}_8$ have multiplicative inverses?

To solve an equation such as $3x = 4 \pmod 5$, we multiply both sides by the inverse of $3 \pmod 5$, which is 2. This yields

$x = 3 \pmod 5$. But what are we to do when faced with the equation $4x = 7 \pmod{10}$, if $4^{-1} \pmod{10}$ doesn't exist? Two things could happen; either there is no solution, or there is more than one. Which occurs depends on the numbers involved in the equation. There is no solution for $4x = 7 \pmod{10}$, but $4x = 6 \pmod{10}$ has solutions of $x = 4$ and $x = 9$. In other situations, there may be more solutions. For example, $13x = 13 \pmod{26}$ has 13 different solutions. (Can you find them all?)

> ✓ Solve $3x = 5 \pmod 7$ and $3x = 5 \pmod 8$.

When a multiplicative inverse does exist, there is no easy formula for finding it as there was for finding additive inverses. (We will be able to compute multiplicative inverses using a theorem given later in the chapter. However, trial-and-error may be your best strategy for a small modulus.)

More Cryptology

The shift cipher was too simple to be secure. Another method along the same lines as the shift cipher is to take the plaintext letter p (actually the residue mod 26 associated with the letter) and multiply it before shifting. That is, the formula for encryption is $p \cdot m + s = c \pmod{26}$, where m is a multiplier and s is a shift. For example, let's take a multiplier of 5 and a shift of 18, giving us an encryption formula of $p \cdot 5 + 18 = c \pmod{26}$. Our standard message `MEET ME AT NOON`, or 12 4 4 19 12 4 0 19 13 14 14 13, would be encrypted as 0 12 12 9 0 12 18 9 5 10 10 5, giving the ciphertext `AMMJA MSJFK KF`.

> ✓ What's the encryption formula using the preceding scheme with a multiplier of 3 and a shift of 5? Use it to encrypt the message `THIS MESSAGE IS ENCRYPTED`.

To decrypt this message, the receiver must find the decryption formula (knowing the encryption formula, of course).

That is, the receiver must solve the equation $p \cdot 5 + 18 = c \pmod{26}$ for p. Now $-18 = 8 \pmod{26}$, but we also need to know $5^{-1} \pmod{26}$. We know $5^{-1} \pmod{26}$ exists since $\gcd(5, 26) = 1$. With a little trial and error, we find that $5 \cdot 21 = 105 \equiv 1 \pmod{26}$. This allows us to solve for p:

$$\begin{aligned} p \cdot 5 + 18 &= c \pmod{26} \\ p \cdot 5 + 18 + 8 &= c + 8 \pmod{26} \\ p \cdot 5 \cdot 21 &= (c + 8) \cdot 21 \pmod{26} \\ p &= c \cdot 21 + 12 \pmod{26}. \end{aligned}$$

And so we have the decryption formula. You should check that this really does recover the plaintext from the ciphertext.

> ✓ What is the decryption formula when the encryption formula used a multiplier of 3 and a shift of 5? Use this to decrypt the ciphertext you came up with in the preceding example.

Some care must be taken when selecting the multiplier m. An obvious requirement of any encryption scheme is that once the ciphertext has been produced, you must be able to reproduce the plaintext. In the scheme we've just described, this means we must be able to solve for p. But this is only possible if $m^{-1} \pmod{26}$ exists, which only happens when $\gcd(m, 26) = 1$. Thus the only multipliers we can legally use here are 1, 3, 5, 7, 9, 11, 15, 17, 19, 21, 23, and 25. These 12 multipliers and 26 shifts give us $12 \cdot 26 = 312$ different encryption schemes. (Actually, only 311 since a multiplier of 1 and shift of 0 gives us the plaintext as the ciphertext, which isn't what we want.) This is much better than using only shifts, but still not very secure when we could easily try all possible encryptions with the aid of a computer.

To illustrate what can go wrong when using a multiplier that doesn't have an inverse mod 26, consider the encryption scheme $p \cdot 4 + 5 = c \pmod{26}$. (The shift could be any value here.) For instance, plaintext letters `D` and `Q` both get mapped to ciphertext letter `M`. So if the receiver is faced with decrypting an `M`, there is a choice to be made, which is not what you want when decrypting a message. Indeed, this happens for every possible ciphertext letter. As you can check, the only letters that can be ciphertext letters (that is, the only letters that plaintext letters get transformed to) are `A`, `C`, `E`, `G`, `I`, `K`, `M`, `O`, `Q`, `S`, `U`, `W`, and `Y`; those with even values.

Sometimes three extra characters (blank, period, and maybe a comma) are introduced to our alphabet so we do our arithmetic in $\mathbb{Z}_{29}$. Since 29 is prime, all nonzero multipliers will have inverses, eliminating the problem just encountered.

Fermat's Little Theorem

We now give a theorem that will be useful to us, in a more general form, in RSA encryption. It is about 300 years old.

Fermat's Little Theorem. *If p is prime and $\gcd(a, p) = 1$, then*

$$a^{p-1} \equiv 1 \pmod{p}.$$

We will not prove this theorem here, but refer you to any introductory book on number theory; you should find the proof accessible. If we multiply both sides of the congruence by a, we get $a^p \equiv a \pmod{p}$. Now if $\gcd(a, p) > 1$, then a must be a multiple of p since p is prime. But then $a \equiv 0 \pmod{p}$ and so $a^n \equiv 0 \pmod{p}$ for any power n. In particular, $a^p \equiv a \pmod{p}$. Thus this last congruence holds regardless whether or not a and p are relatively prime. Fermat's Little Theorem is sometimes stated in this form.

✓ Compute $a^{p-1} \pmod{p}$ for $a = 3$ and $p = 5$ and also for $a = 4$ and $p = 7$ to verify that you do indeed get 1. Check Fermat's Little Theorem with other appropriate values.

We will use Fermat's Little Theorem as a test to see if a number is composite. If we want to know if n is composite, we choose a number a with $\gcd(a, n) = 1$ and compute $a^{n-1} \pmod{n}$. If this number is not 1, then n is composite. For if it were prime, Fermat's Little Theorem says the result must be 1. Of course, if the result is 1, n may be prime or it may be composite; the result of this test is inconclusive. (This is the theorem we referred to earlier when we said you could tell a number was composite but not know its factors.) If the result of this calculation is 1, we usually keep trying for different values of a. As a matter of convenience, we usually let $a = 2$ to start with and then increment a, if the result of the computation is 1. If you find an a where the computation is not 1, you stop and declare the number composite. This test is usually performed on large n, so it would not be practical to try *all* $a < n$. As a probabalistic primality test, you stop after so many computations and declare n a "probable prime" if all your choices of a yielded $a^{n-1} \pmod{n} = 1$. For large numbers, this is not a practical test to perform with paper and pencil.

> ✓ Calculate $a^{n-1} \pmod{n}$ for all $a < n$ where $\gcd(a, n) = 1$ and $n = 9$. Do you conclude that 9 is composite? Now test for $n = 7$.

You might think that if you calculated $a^{n-1} \pmod{n}$ for all $a < n$ where $\gcd(a, n) = 1$ and you got 1 each time, then n would be prime. Indeed, the converse is true, by Fermat's Little Theorem. You'd *almost* be correct. Unfortunately, there are a few such numbers that are composite, called *Carmichael numbers*, but they are rare. The three smallest Carmichael numbers are 561, 1105, and 1729. Indeed, there are only 105,212 Carmichael numbers less than 10^{15}. (See *Mathematics of Computation*, 61:203 (July 1993), pp. 381–391.) Thus if the preceding test ever yields a calculation where $a^{n-1} \pmod{n} \neq 1$, you know with certainty that n is composite. If you never find such an a, you can only suspect n is prime. That's why we call this a test for composite numbers and not a primality test.

Fast Exponentiation

This test of composite numbers is all well and good, but what if the n you're checking is quite large? To compute $a^{n-1} \pmod{n}$ using the straightforward method of repeated multiplication would take an inordinate amount of time. Fortunately, there is an efficient method for computing any power of a number, appropriately called fast exponentiation. To get an idea for how this works, let's compute 3^{11}. (We'll do our arithmetic in the natural numbers and consider the modulus later.) We'll cleverly write the exponent 11 as a sum of powers of 2: $11 = 1 + 2 + 8$. We will compute 3^{11} by doing as few multiplications as possible. Note that $3^{1+2+8} = 3^1 \cdot 3^2 \cdot 3^8$. We'll call these powers of 2 on the right the "desired factors of 3^{11}" in the discussion that follows. To compute 3^{11}, we will computer higher and higher powers of 3 (the base). We start with 3^1 and since this power of 3 is a one of the desired factors of 3^{11}, we multiply our running total (initially equal to one) by 3^1. The next power of 3 we might need is $3^2 = 9$, which we get by squaring 3^1. This power of 3 is also a desired factor of 3^{11}, so we multiply our running total by this amount, getting 27. We calculate the next power of 3 we might need, $3^4 = 81$, by squaring 3^2. But 3^4 is not a desired factor of 3^{11}, so we don't change our running total. The next power of 3 we might need is $3^8 = 6561$, which is a desired factor, and so we multiply our running total by 6561 to get 177147. This ends our computation. Note that we needed only six multiplications to compute 3^{11}. Looping through repeatedly multiplying by 3 would take 10 multiplications. But this is a small exponent. The savings are dramatic for large exponents, as we will see.

The remaining problem is to find what we called the desired factors. We find the desired factors by expressing the exponent as a sum of powers of 2. The algorithm for fast exponentiation is given next. Notice how the desired factors are picked off. Recall, $\lfloor x \rfloor$ is the floor function (the greatest integer no larger than x).

```
function fastexp(a,b)
% require:  a & b positive integers
% ensure:  function returns a^b
exp = b;
total = 1;
factor = a;
while (0 < exp) do
  % do we have a "desired factor"?
  if (exp mod 2 = 1) then
    total = total * factor;
  endif;
  exp = ⌊exp/2⌋;
  % square the factor
  factor = factor * factor;
endwhile;
return(total);
end function;
```

Let's trace through the calculation 5^{23} using `fastexp`. We'll trace the values of `exp`, `total`, and `factor` *after* each pass through the `while` loop.

	exp	total	factor
initially:	23	1	5
	11	5	25
	5	125	625
	2	78125	390625
	1	78125	152587890625
	0	11920928955078125	*****

The last value of `total` is what `fastexp` returns for the value of 5^{23}. We do not show the last value of `factor` since it is not used in any subsequent calculation and it is 23 digits long! You should notice that since `factor` is squared each time through the loop it grows at an increasingly rapid rate. This is a good example of *exponential explosion.*

During each iteration of the loop, we take the floor of half the exponent. The loop stops when this value gets to 0. Therefore, the number of times through the loop is about $\log_2 b$, where b is the exponent, as opposed to about b repetitions for the straightforward algorithm of repeated multiplication. For example, if $b = 1000$, fast exponentiation would loop 10 times as opposed to 999 times for the straightforward method. Indeed, this algorithm deserves to be called fast.

> ✓ Compare the number of iterations of fast exponentiation with the straightforward method if b =1,000,000.

If we wish to compute $a^b \pmod{n}$, which we will want to do for our encryption examples later on, we modify the fast exponentiation algorithm to perform a "mod by n" operation each time a multiplication is done; so two lines would need to be modified. The size of the exponent that's used in practice is on the order of 200 digits! We certainly need fast exponentiation in this case.

> ✓ Make the necessary modification to `fastexp` to compute $a^b \pmod{n}$.

> ✓ Now use your modified `fastexp` to compute $3^{50} \pmod{10}$.

Euler's Theorem

Euler later extended Fermat's Little Theorem, introducing a new function. This function seems a little off-beat and indeed it is hard at first to see how it would be useful or why anyone would be interested in the function to begin with. It illustrates Euler's genius.

Euler's *phi function*, written $\varphi(n)$ and defined for all natural numbers, is the number of natural numbers less than n that are relatively prime to n. (This function is sometimes called the totient function.) This function counts how many of those numbers there are; it has nothing to do with what values those numbers have. Since every number is relatively prime to 1, $\varphi(n) \geq 1$. For example, $\varphi(10) = 4$ since 1, 3, 7, and 9 are relatively prime to 10. Also, note that $\varphi(8) = 4$, $\varphi(11) = 10$, and $\varphi(15) = 8$.

There are three facts that will aid in computing $\varphi(n)$. First, if n is prime, then $\varphi(n) = n - 1$, since all numbers less than n are relatively prime to n. The two other facts used in calculating $\varphi(n)$ are not so obvious and we will only state them: (1) If $\gcd(n, m) = 1$, then $\varphi(n \cdot m) = \varphi(n) \cdot \varphi(m)$, and (2) if p is prime, then $\varphi(p^n) = p^n - p^{n-1}$. These, along with the Fundamental Theorem of Arithmetic, allow us in principal to compute $\varphi(n)$ for all natural numbers; we first factor n into a product of powers of primes, then use the three facts about the phi function we just mentioned. For example, let's compute $\varphi(1200)$. First, we factor 1200: $1200 = 2^4 \cdot 3 \cdot 5^2$. But then

$$\begin{aligned} \varphi(1200) &= \varphi(2^4)\,\varphi(3)\,\varphi(5^2) \\ &= (2^4 - 2^3)\,2\,(5^2 - 5) \\ &= 320. \end{aligned}$$

Note that the first step was due to rule (1), which we could apply because the three factors listed were relatively prime. You see that this method of determining $\varphi(n)$ requires that we be able to factor n, which is difficult for large n.

✓ Calculate $\varphi(100)$ and $\varphi(40)$.

We can now state Euler's theorem. This is a generalization of Fermat's Little Theorem (that is, Fermat's Little Theorem is a special case of Euler's theorem) since if n is prime, then $\varphi(n) = n - 1$.

Euler's Theorem. *If* $\gcd(a, n) = 1$, *then* $a^{\varphi(n)} \equiv 1 \pmod{n}$.

For example, if $n = 10$, $\varphi(10) = 4$ so for every a such that $\gcd(a, 10) = 1$, $a^4 \equiv 1 \pmod{10}$. Try this with a couple of values for a. While this example is easy to calculate without this theorem, calculating some larger powers is not so easy. For instance, $49^{320} \pmod{1200}$ would take some tedious calculations by hand, even with fast exponentiation, but Euler's theorem tells us the value is 1. (Don't forget the requirement that 49 and 1200 be relatively prime.) Euler's Theorem can also be used as a shortcut for other calculations. For example, when calculating $7^{322} \pmod{1200}$, note that $7^{322} = 7^{320}\, 7^2 \equiv 7^2 \pmod{1200} \equiv 49 \pmod{1200}$.

> ✓ Use Euler's theorem as an aid to speed the calculating of $7^{82} \pmod{100}$.

RSA Encryption

We now have all the pieces in place to describe RSA encryption. Before getting down to the details, we'll give an overview of the method. RSA encryption is an example of *public-key encryption*, which means that the encryption key is made public and the decryption key is held private. That is, the receiver (whom we'll call person A) makes known to everyone the key to encrypt messages to person A. Despite knowing this key, it is very unlikely that the senders will be able to figure out the decryption key, which person A keeps private. If there are several people using this method, then each has his or her own keys. Keep in mind that you use the public encryption key of the person to whom you are sending the message. (The receiver will, of course, know his or her own decryption key.)

The first step in RSA encryption is to translate the message into a number M. (There are some restrictions on the size of M.) The encryption key is a pair of numbers e and n. The ciphertext C is computed by $M^e \bmod n = C$. The receiver recovers the plaintext M by applying the decryption key (again a pair d and n): $C^d \bmod n$. This value will be equal to M, because of our choices of e, d, and n, which we describe next.

Let's go through the steps the receiver needs to perform in order to find appropriate values e, d, and n. We'll use some very small numbers in our example to make the arithmetic easier. In practice, these numbers need to be quite large, in order to make this method is secure.

First, two primes p and q are picked. (Let's pick $p = 7$ and $q = 11$.) These are kept secret. Indeed, if anyone could figure out these two primes, the cipher is easily cracked. In practice, p and q each have about 100 digits each. Compute $n = p \cdot q$. (Here, $n = 77$.) We now need $\varphi(n)$. (In our example, $\varphi(77) = 6 \cdot 10 = 60$.)

Now we find a number e that is relatively prime to $\varphi(n)$; there are many choices for e. (Let's pick $e = 53$.) The encryption key, which is made public, is the pair e and n. (So $e = 53$ and $n = 77$.)

✓ Suppose you pick primes 5 and 7. Find an appropriate value for e. There are many choices.

The next step is to calculate d, which will be the multiplicative inverse of e mod $\varphi(n)$. We use the extended Euclid's algorithm to do this: Since $\gcd(\varphi(n), e) = 1$, we can find a and b so that $1 = e \cdot a + \varphi(n) \cdot b$. Thus, $1 - \varphi(n) \cdot b = e \cdot a$. In other words, $e \cdot a = 1 \mod (\varphi(n))$; that is, a is the multiplicative inverse of e. Thus this number a that we get from the extended Euclid's algorithm is the value for d we want. The decryption key is then the pair d and n. (In our example $d = 17$, since $53 \cdot 17 = 1 \pmod{60}$.)

✓ Using primes 5 and 7 and your aforementioned choice of e, calculate d.

To encrypt a message, we first translate into a number. We use some code to translate the characters to two-digit numbers.

Let's slightly change our scheme from before and let `A`= 01, ... `Z`= 26, space = 27, and period = 28. Then we concatenate the numbers to form a long string of digits. For example, suppose or message was, again, `MEET ME AT NOON.` Then the encoded message would be 1305052013050120141515 1428. Since the number encrypted must be less than the n of the keys, we may need to break this long string of numbers into blocks of numbers each less than n. For instance, if $n = 7001940587$, we could split our message up into blocks 1305052013, 0501201415, and 151428, and encrypt and send each block separately. The receiver must then decrypt each block and reassemble them to form the message. (In our example, n was unrealistically small; $n = 77$. So our blocks would each be less than 77, which amounts to one character. To encrypt, we take the block, say $M = 13$, and encrypt according to the encryption formula given previously: $13^{53} \pmod{77} = 41$. To decrypt, $41^{17} \pmod{77}$ and you'll see this is 13, the original message.)

> ✓ For primes 5 and 7 and your choices for e and d, encrypt the "message" 10. Then decrypt your result. You should, of course, recover 10.

For more realistic examples, a computer algebra system, like *Maple* or *Mathematica*, should be used. (Or, write your own programs to handle large integer arithmetic.)

Let's go through a small example, but one too large to do by hand. We use *Maple*, and go through the same steps we did in our first very small example. First we pick two primes, let's say $p = 299011$ and $q = 23417$. Then $n = 7001940587$ and $\varphi(n) = (p-1)(q-1) = 7001618160$. We then find a number e relatively prime to $\varphi(n)$; we pick $e = 234569$. So we have an encryption key of $e = 234569$ and $n = 7001940587$. We find, using the extended Euclid's algorithm, the multiplicative inverse of e mod $\varphi(n)$ and assign that to d. We find that $d = 171042949$ and so our decryption key is $d = 171042949$ and $n = 7001940587$.

Our message `MEET ME AT NOON` was changed to numbers and broken into three blocks, since the blocks must each be less than n: 1305052013, 0501201415, and 151428. We then encrypt each block, using our encryption key (e, n), and get ciphertext blocks 5210600508, 2476435985, and 5003991730. These are the ciphertexts that are sent. The receiver, who is the only one to hold the decryption key, knows that $d = 171042949$, and decrypts each block to recover the three original plaintext blocks. Note that the second block has a leading 0, which we don't usually write. The receiver knows that there should be a leading 0, however, since each letter corresponds to two digits and a plaintext block with an odd number of digits occurred after decryption. You should try out this example on your own computer algebra system to see that you get the same numbers.

Why does this work? That is, why does decrypting the ciphertext guarantee the result is the original plaintext? Recall that we picked e and d so that $e{\cdot}d \pmod{\varphi(n)} = 1$. That is, the remainder when dividing $e \cdot d$ by $\varphi(n)$ is 1. So by the Division Theorem, there is a number q so that $e \cdot d = q \cdot \varphi(n) + 1$. Now, let M be the plaintext. If we encrypt M, then decrypt it, we are calculating

$$\begin{aligned} M^{e\cdot d} \pmod{n} &= M^{q\cdot\varphi(n)+1} \pmod{n} \\ &= M^{q\cdot\varphi(n)} \cdot M \pmod{n} \\ &= (M^{\varphi(n)})^q \cdot M \pmod{n}. \end{aligned}$$

But Euler's theorem says that $M^{\varphi(n)} = 1 \pmod{n}$, so this last expression is $1 \cdot M = M \pmod{n}$. Thus, the plaintext is recovered.

Exercises

1. Show that the relation $\sim_n$ is an equivalence relation on the set of integers. Recall that $a \sim_n b$ iff $n|(a - b)$.

2. Find q and r (the quotient and remainder) guaranteed by the Division Theorem for $a = 100$ and $b = 57$. For $a = 407$ and $b = 10$. For $a = 10$ and $b = 407$.
3. Use Euclid's algorithm to find $\gcd(1001, 533)$, $\gcd(330, 204)$, and $\gcd(1000000, 7322)$.
4. Use the calculations you performed in the last exercise to find the x and y so that $\gcd(a, b) = ax + by$ for the various a and b given.
5. Suppose you use the sieve of Erastothenes to find all primes less than 1000. (Don't do this.) You repeat the main loop how many times? Same question for 1,000,000.
6. Show, using induction, that if p is prime and $p|a_1a_2\cdots a_n$, then $p|a_i$ for some i, $1 \le i \le n$.
7. Using the sieve of Eratosthenes, find all two-digit primes.
8. Using the function Prime, given in the text, determine if 563 is prime or not. How many iterations of the loop in this function did you perform? Now determine if 389 is prime or not.
9. Solve each of the following:

$$\begin{aligned} 3 + x &= 7 \pmod 8 \\ 4 + x &= 3 \pmod{10} \\ 137 + x &= 100 \pmod{903}. \end{aligned}$$

10. Solve each of the following:

$$\begin{aligned} 2x + 3 &= 1 \pmod 7 \\ 4x + 7 &= x + 2 \pmod{10} \\ 5x + 10 &= 2x + 7 \pmod{20}. \end{aligned}$$

11. List all solutions to $2x = 4 \pmod{12}$. How many solutions are there to $2x = 5 \pmod{12}$?

12. In the function `Prime`, we only checked if a divides n for $a \leq \sqrt{n}$. Why does this suffice?
13. Find the additive inverses of all the elements of $\mathbb{Z}_{10}$. Same for $\mathbb{Z}_9$.
14. Find the multiplicative inverses for all numbers that have them in $\mathbb{Z}_7$, in $\mathbb{Z}_{10}$, in $\mathbb{Z}_{11}$, in $\mathbb{Z}_{12}$, in $\mathbb{Z}_{15}$ and in $\mathbb{Z}_{20}$.
15. Factor completely into primes: 30, 90, 100, 101, 1000 and 1002.
16. Find $\varphi(n)$ for $n = 30, 71, 99, 999, 10^6$ and 97.
17. Prove that $\gcd(m,n) \cdot \operatorname{lcm}(m,n) = mn$. ($\operatorname{lcm}(n,n)$ is the least common multiple of n and m.)
18. Show that $\frac{(3^{77}-1)}{2}$ is odd and composite. *Hint*: Find $3^{77} \bmod 4$.
19. Prove or disprove: $\gcd(km, kn) = k \gcd(m,n)$.
20. Prove or disprove: $\operatorname{lcm}(km, kn) = k \operatorname{lcm}(m,n)$.
21. Prove that a number is divisible by 3 iff the sum of its digits is divisible by 3.
22. Prove that a number is divisible by 4 iff the number defined by the 2 rightmost digits is divisible by 4.
23. Prove that a number is divisible by 9 iff the sum of its digits is divisible by 9.
24. As illustrated in our discussion of RSA encryption, we can calculate $a^{-1} \bmod n$ by using the extended Euclid's algorithm (provided a has an inverse mod n). Use this method to calculate the following: $5^{-1} \bmod 7$, $9^{-1} \bmod 100$, and $11^{-1} \bmod 26$.
25. Use fast exponentiation to calculate $12^{320} \bmod 20$ and $5^{210} \bmod 12$.
26. What is the encryption formula for encrypting a shift cipher when plaintext A is encrypted as E? What is the decryption formula?

27. Decrypt the ciphertext `ZWTSW GOAMG HSFM` if the shift cipher was used with a shift of 14. What is the decryption key?
28. Encrypt the message `HELP IS ON THE WAY` using the encryption formula $3p+11 = c \bmod 26$. Find the decryption formula and decrypt the ciphertext you got. (It should be the original plaintext, of course.)
29. Find the decryption formula for the encryption formula $7p + 20 = c \bmod 26$. For $21p + 4 = c \bmod 26$.
30. Encrypt all 26 letters using the formula $13p = c \bmod 26$. Now encrypt all 26 letters using $2p = c \bmod 26$. Why do both of these multipliers pose a problem for decryption?
31. When using RSA encryption, suppose you start with the small primes 5 and 11. Find appropriate values for e and d. What's the largest number you can encrypt with your values?
32. When using RSA encryption, suppose you start with the small primes 13 and 19. Find appropriate values for e and d. What's the largest number you can encrypt with your values?
33. If you have access to a computer algebra system, such as *Maple* or *Mathematica*, develop your own RSA encryption key. Start by finding two primes of about 10 digits each. This would make your message blocks about 20 digits long. To be more practical, start with 50-digit primes. Test your keys by encrypting and decrypting a short message.

Programming Problems

1. Write a program that inputs a positive integer n and outputs the prime factorization of n. For example, if $n = 100$, the output should be 2, 2, 5, 5. If $n = 101$, the output should be 101.
2. Write a program that prints *all* divisors of n.

3. Write a program that inputs a positive integer n and prints $\varphi(n)$. You'll need to find the prime factorization of n.
4. Write a program that inputs a, b and n and prints $a^b \bmod n$. Use fast exponentiation in your calculation.
5. Write a program that inputs positive integers n and m and prints $\gcd(n, n)$.
6. Using the gcd function you wrote in the last programming problem, write a program that computes $\text{lcm}(n, m)$. ($\text{lcm}(n, n)$ is the least common multiple of n and m.)
7. Write a program that inputs positive integer n and prints all positive integers $< n$ that have multiplicative inverses mod n, along with their inverses. (Assume n is small enough that you can find inverses by exhaustive search. Later, in Chapter 6, we give an algorithm for the extended Euclid's algorithm.)
8. Write a program that solves congruences: Input a, b, c, and n and prints the solution (if one exists) for the $ax+b = c \mod n$.
9. Write a program that encrypts messages using an encryption formula of $mp + s = c \bmod 26$, where the user supplies the values for m and s, making sure that m is a legal multiplier. After entering m and s, have the user enter a message. Ignore all nonalphabetic characters and print your ciphertext in groups of five letters.
10. Write a program that prints all primes less than n, which is entered by the user, using the sieve of Eratosthenes. Assume $n < 100000$.
11. Write a program that adds fractions. The user should enter the numerator and denominator (both positive integers) of two fractions. Output should be the sum of these two fractions in reduced form. (You will need the gcd function to reduce your sum to lowest terms.)
12. Write a program that produces a table with two columns: n and $\varphi(n)$.

13. Write a program that produces a table with two columns: n and the factors of n.
14. A composite number n that satisfies $b^{n-1} \equiv 1 \mod n$ for every positive integer b such that $\gcd(b, n) = 1$ is called a *Carmichael* number. Write a program that produces Carmichael numbers. It's interesting to see how high a Carmichael number you can generate within the bounds of the integers allowed by your programming language.
15. The EKG sequence is a sequence of positive integers generated as follows: The first two numbers of the sequence are 1 and 2. Each successive entry is the smallest positive integer not already used that shares a factor with the preceding term. So, the third entry in the sequence is 4 (being the smallest even number not yet used). The next number is 6 and the next is 3. The first few numbers of this sequence are given below.

    $1, 2, 4, 6, 3, 9, 12, 8, 10, 5, 15, 18, 14, 7, 21, 24, 16, 20, 22$

    While the sequence has a general increasing trend, it gets its name from its rather erratic fluctuations. Write a program that prints the first n entries of the EKG sequence.

Chapter 6
Recursion

You are familiar with procedures and functions from your beginning programming course. Procedures and functions can be called by other procedures and functions. Most modern programming languages allow procedures or functions to call themselves. In this case we say that the procedure or function is *recursive*. Recursion is a very powerful technique for designing algorithms and it has many important applications. Recursion is an idea closely related to induction. Indeed, the words inductively and recursively can be interchanged when pertaining to the method of constructing or calculating something. We will only touch on some of the simpler examples.

Recursion has its good and bad points. In some cases, a recursive procedure is much simpler to write than an iterative version of the same. Moreover, when this happens it is often easy to convince yourself that the program is correct. On the other hand, debugging a recursive procedure sometimes takes a little more patience than a nonrecursive one. Moreover, it frequently takes a lot of thought initially to design, if you're not in the habit of thinking recursively.

From the implementation point of view, there is an efficiency question to consider. Any task accomplished by a recursive program can also be done iteratively. Sometimes one approach seems more natural than the other. In any case, the recursive function or procedure requires more overhead than its iterative counterpart during execution and results in a slower program. When you study algorithm analysis, you will find precise methods for comparing one algorithm to another.

Here, we will present some examples of recursion in order to add this approach to our collection of design techniques for algorithms. As a first example of recursion, let's consider writing a function to compute $n!$, for nonnegative integers n. While it is easy to write a function that computes $n!$ iteratively, the definition for $n!$ is usually given recursively:

$$n! = \begin{cases} 1, & \text{if } n = 0 \text{ or } n = 1 \\ n \cdot (n-1)!, & \text{otherwise.} \end{cases}$$

This is the form that recursive definitions take. First, a base case is presented. Here, this occurs when n is 0 or 1; the value of $n!$ is 1 in this case. The remaining cases are when $n > 1$ and are defined in terms of factorial of a smaller value of n. You should notice a strong similarity with induction in this setup.

Writing a recursive function for it is a matter of translating the mathematical definition into code:

```
function factorial(n)
  % require:  n is a non-negative integer
  % ensure:  function returns n!
if n < 2 then                   % base case
   return(1)
else
   return(n * factorial(n-1))
endif
end function
```

Here, the program is exactly the mathematical definition and so we can assume the program is correct as long as the syntax is correct. The iterative version of this function is probably familiar to you. We show the iterative version next for comparison. In this case there is little difference between the two implementations when it comes to understanding them.

```
function factorial-iterative(n)
  % require:  n is a non-negative integer
  % ensure:  function returns n!
fact=1
for i=2 to n do
```

```
      fact = fact * i
    endfor
    return(fact)
    end function
```

The correctness of a recursive function is easy to see mathematically but it is not so obvious how this computation is done on a computer. Let's suppose you make the call `factorial(10)`. Note that the value returned for this call will be $10 \cdot$ `factorial`(9). To compute the value of `factorial(9)` requires another recursive call, so the final computation (the multiplication of 10 and whatever value `factorial(9)` returns) must wait until the call to `factorial(9)` returns with its value. But then, `factorial(9)` requires the value of `factorial(8)`, which requires the value of `factorial(7)`, which requires the value of ... you get the idea. These recursive calls keep going until the base case `factorial(1)`, which returns 1. Now, this returned value can be used to compute `factorial(2)`, which returns its value so `factorial(3)` can be computed, and so on until `factorial(9)` returns with its value and the value of `factorial(10)` can be computed and returned.

> ✓ Trace through all the recursive calls for initial call `factorial(6)`. What value do you get? How many recursive calls are made (not counting the initial call)?

This is a simple example of recursion. There are more complicated recursive functions. All recursive functions have two important requirements: A base case (or base cases) must be provided, and the recursive calls (to "simpler" instances of the computation) must be handled properly to compute the value you wish. This means that the parameter(s) to the recursive calls must be correct and that the resulting value of the recursive call must be used correctly.

We emphasize the necessity of the base case: In a recursive function it is critical that we have a condition (called the base case) that returns a value computed nonrecursively and that *this condition will always occur eventually regardless of what the initial value of the parameter is.* (Mathematically, this means the function is well-defined.) If this condition is not met, infinite recursion may occur and the function may never return (until storage space for the call frame runs out; then you'll have abnormal termination). Here's an example of a recursive function that will not always return. In this case, it doesn't return when $n < m$. (Don't try to "fix" this function; there was nothing useful intended.)

```
function oops(n,m)
if (n>m) then
  return(m)
else
  return(oops(n-1,m) + 1)
end function
```

> ✓ Explain why this function fails to return when $n < m$.

Let's look at a more complicated function, one that computes the nth Fibonacci number. When you see the mathematical definition of the function, you'll see that it isn't any more difficult to program than the factorial function. Recall that the Fibonacci numbers form the sequence $1, 1, 2, 3, 5, 8, 13, 21, \ldots$. In general, you compute the next Fibonacci number by summing the previous two. Of course, we need to provide a starting point for this sequence, and we start with the 1 and 1 as the first two elements of the sequence. Here's the mathematical formula for the nth Fibonacci number:

$$fib(n) = \begin{cases} 1, & \text{if } n = 1 \text{ or } n = 2 \\ fib(n-1) + fib(n-2), & \text{otherwise.} \end{cases}$$

The code for this function is now just the straightforward translation of this description:

```
function fib(n)
  % require:  n is a positive integer
  % ensure:  function returns the nth
  %          Fibonacci number
if n=1 or n=2 then
  return(1)
else
  return(fib(n-1) + fib(n-2))
endif
end function
```

Tracing back through the calls for this function is more complicated than for factorial since the computation requires two recursive calls. Thus the total number of calls required to compute `fib(n)` approximately doubles as n increases by one. Let's trace through the calls if the initial call is `fib(4)`.

The value returned for `fib(4)` is the sum of `fib(3)` and `fib(2)`. Thus two recursive calls must be made. While it makes no difference in the calculation, let's assume that the one listed first (that's `fib(3)` here) is called first. The call to `fib(2)` is not made until `fib(3)` returns. Now the call to `fib(3)` returns the sum of `fib(2)` and `fib(1)`. Thus, there are two more recursive calls. The calls to `fib(2)` and `fib(1)` (three calls in total) are base cases and so each simply returns 1 to the calling function. This makes a total of four recursive calls. (We don't count the initial call here.) We can illustrate the calls in the following tree diagram. The calling functions are above the called functions:

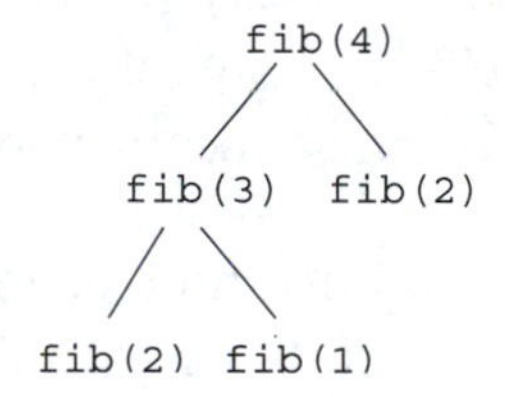

✓ Trace through a call to `fib(5)`. Draw a diagram of the recursive calls made. How many are there? How many duplicate calls were made? Repeat for `fib(6)`.

An important point here is that the actual calls and computation may be complicated, but the writing of the function isn't. We can let the built-in function calling mechanisms of the computer (and the language you're using) do the work for us. As an exercise at the end of the chapter, you are asked to write an iterative version of `fib` and compare it to the version given previously.

Binary Search

One of the most frequent tasks that computers perform is that of searching a list for an item. The item searched for might be a complicated structure, but typically we search for an item (or all items) with a value in a certain field. This value is called the search *key.* If the list is sorted and stored in an array, we can efficiently find an item with a given key value, if it exists, or determine that no such item exists. The idea is simple and is similar to the strategy used in the children's guessing game where you try to guess a number between 1 and 100 (or between any two bounds). After each guess, the holder of the secret number responds with either "correct," "too high," or "too low." You quickly figure out that the best strategy is to guess in the middle and, depending on the answer you get, adjust your range upward or downward (if incorrect) and repeat the process until you guess correctly. Note that this scheme is recursive because you repeat the process with different parameters. Here the parameters are the lowest and highest possible values for the unknown number.

✓ Using the aforementioned scheme and starting with a range from 1 to 100, what numbers would you guess if the secret number were 18?

> ✓ If the range were initially 1 through 1000, how many guesses would you make in the *worst case* if you used this scheme? What if the range were 1 through 1,000,000?

Our problem here is to search for the *index* of a particular element in an array. It is assumed that the array is sorted in nondecreasing order. This function will return an index where a given value occurs (if it occurs at more than one index it will return only one of those indices) or return a value of −1 if it is not in the array. (We'll assume indexing starts after −1.) This algorithm is called *binary search*:

```
function bin-search(A, low, high, value)
  % require:  A[low..high] is an array in
  %           non-decreasing order, value is
  %           of the same type as the array
  % ensure:   function returns the index i, so
  %          that low ≤ i ≤ high
  %          and A[i] = value
% searching an empty list?,
%  if so, then not there
if (low > high) then
  return(-1)
else
  mid = ⌊(low + high)/2⌋
  if (A[mid] = value) then
    return(mid)
  else
    % value in low half of list?
    if (A[mid] > value) then
      return(bin-search(A, low, mid-1, value))
    else  % value in high half of list
      return(bin-search(A, mid+1, high, value))
    endif
  endif
```

```
endif
end function
```

As an example, let's suppose the array A is indexed from 1 through 30 and the elements of the array are

$$3, 4, 7, 9, 10, 11, 12, 13, 15, 17, 19, 20, 23, 25, 26,$$
$$27, 28, 30, 33, 34, 37, 38, 40, 42, 50, 51, 52, 53, 54, 56.$$

Suppose we are searching for a key value of 17. Thus our initial call is `bin-search(A,1,30,17)`. The value for `mid` is 15, and since `A[15]=26` is greater than 17, the next call is `bin-search(A,1,14,17)`. The next value for `mid` is 7, and `A[7]=12` is less than 17. So the next call is `bin-search(A,8,14,17)`. The subsequent calls would be `bin-search(A,8,10,17)` and `bin-search(A,10,10,17)`. But `A[10]=17`, which is the key value we are searching for, and so 10 is returned to the calling function, which returns the 10 to its calling function and so on until 10 is returned from the initial call.

> ✓ Trace through the recursive calls if the initial call is `bin-search(A,1,30,23)`. Now trace through the calls if the initial call is `bin-search(A,1,30,24)`. Note that in the latter case, a -1 should be returned since 24 is not in the array.

Compare this with the linear search, that starts by comparing the given value with the first element of the list and then proceeds to the second element and so on. The linear search may have to examine every element of the list while the binary search at each step cuts in half the number of elements it needs to examine. Even without a very careful analysis of these algorithms, we see that binary search is superior.

Euclid's Algorithm

In Chapter 5 we used Euclid's algorithm to find the gcd of two numbers. The algorithm was presented in an iterative manner, but it is also possible to think about this algorithm recursively. This is true of many algorithms where you repeat the same steps but each iteration uses values computed in the previous step.

```
function gcd(n,m)
  % require:  n and m are non-negative
  %            integers not both zero
  % ensure:   function returns the gcd
  %           of n and m
if (m = 0) then
  return(n)
else
  return(gcd(m, n mod m))
endif
end function
```

Let's trace through the calls if the initial call is `gcd(12,63)`. This call would call `gcd(63,12)`, which would call `gcd(12,3)`, which would call `gcd(3,0)`, which would return 3. Compare this with the iterative version from the previous chapter. You should see that nearly the same calculations are taking place.

> ✓ Trace the execution of `gcd(24,111)`. Do the same for `gcd(10,100)`.

In Chapter 5 we also saw the usefulness of the extended Euclid's algorithm which not only finds the gcd of two numbers n and m but also the numbers a and b so that $\gcd(n, m) = a \cdot n + b \cdot m$. In the simple Euclid's algorithm, we went from computing $\gcd(n, m)$ to computing $\gcd(m, n \bmod m)$. Suppose $d = \gcd(m, n \bmod m)$. (Of course, $d = \gcd(n, m)$ also.) The problem here is, if we have found (recursively) values for a' and b' so that $d = a' \cdot m + b' \cdot (n \bmod m)$, we need to compute a and b so that $d = a \cdot n + b \cdot m$.

Suppose we know not only $d = \gcd(m, n \bmod m)$ but also

a' and b' so that

$$d = a' \cdot m + b' \cdot (n \bmod m).$$

We now need to see if from these values we can find the corresponding values for a and b. Of course, the answer is yes (or why would we bring it up?). First we note that

$$n \bmod m = n - \lfloor n/m \rfloor \cdot m,$$

and by substituting see that

$$\begin{aligned} d &= a' \cdot m + b' \cdot (n - \lfloor n/m \rfloor \cdot m) \\ &= b' \cdot n + (a' - b' \cdot \lfloor n/m \rfloor) \cdot m. \end{aligned}$$

So we have that $a = b'$ and $b = a' - b' \cdot \lfloor n/m \rfloor$. We can now give a procedure for the extended Euclid's algorithm. Note that the gcd is passed back in the parameter list here, unlike the plain gcd function, which returns the value.

```
procedure gcdext(n, m, d, a, b)
   % require:  n and m are positive
   %          integers
   % ensure:  compute d=gcd(n,m) and a
   %         and b so that d = a*n + b*m
if (m = 0) then
  d = n
  a = 1
  b = 0
else
  gcdext(m, n mod  m, d, a, b)
  c = b
  b = a - ⌊n/m⌋· b
  a = c
endif
end procedure
```

This procedure is a little more difficult than the previous ones. Note that the purpose of a call to `gcdext(n, m,` `d, a, b)` is to find not only `d`, which is the gcd of `n` and `m`, but also find the appropriate values of `a` and `b`. We illustrate by making the call `gcdext(12,63,d,a,b)`. As we've seen before, a series of recursive calls follows: `gcdext(63,12,d,a,b)`, `gcdext(12,3,d,a,b)`, and `gcdext(3,0,d,a,b)`. This last call is a base case and calculates that `d` $= 3$, `a` $= 1$ and `b` $= 0$ and returns to its calling procedure. (We see that, indeed, $3 = 1{\cdot}3+0{\cdot}0$, as we wish.) The calling procedure is `gcdext(12,3,d,a,b)` and this procedure takes these values of `d`, `a`, and `b` and calculates its values of `a` and `b`. (The value of `d` stays the same.) We see that the procedure calculates that `a` $= 0$ and `b` $= 1$. (Again, we see that, indeed, $3 = 0 \cdot 12 + 1 \cdot 3$.) This procedure then returns to its calling procedure, which is `gcdext(63,12,d,a,b)`, and this procedure calculates that `a` $= 1$ and `b` $= -5$. (We see that $3 = 1 \cdot 63 + (-5) \cdot 12$.) These values are returned to the calling procedure, and that procedure calculates that `a`$= -5$ and `b` $= 1$. This is the initial call, and so we see that $3 = \gcd(12, 63)$ and $3 = (-5) \cdot 12 + 1 \cdot 63$. Whew!

✓ Trace through the call `gcdext(24,111,d,a,b)`.

Tower of Hanoi

Finally, on a lighter note, consider the Tower of Hanoi (TOH) game, which is played by moving disks from one peg to another. The game starts with n disks of different diameters stacked from largest on the bottom to smallest on the top on the leftmost of three pegs. The object of the game is to move the disks, one at a time from one disk to another, so that a disk is never placed on top of a smaller disk, until the original stack has been moved to the rightmost peg. (There is the myth of Buddhist monks in heaven solving this puzzle with 100 golden disks on three silver pegs. They make one move a second and the claim is that when they finish the puzzle, the world will end.)

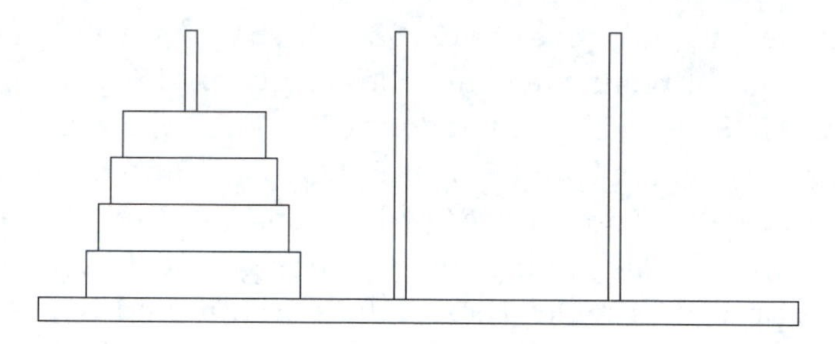

4 disk Tower of Hanoi

We can develop a recursive solution to this game. This game can be solved for n disks in terms of solving the game for $n-1$; reducing a problem to a smaller one is the essence of a recursive solution to any problem. If we have a solution to this smaller problem, then to solve the problem for n disks, we first move the top $n-1$ disks from the left peg to the middle peg (this is the recursion), exposing the largest disk on the left peg. We can then move this disk from the left peg to the right peg. (Note that all the disks smaller than disk n are on the middle peg.) Now we move the $n-1$ disks from the middle peg to the right peg (recursion again). Let's write code for this solution. Our procedure has four parameters: the number of disks, the peg they start on, the peg they wish to move to, and, finally, the other peg, thought of as a temporary storage peg. The output for this procedure will be the instructions for solving this puzzle.

```
procedure TOH(n, start, ending, intermediate)
  % require:  n a positive integer, start,
  %           ending, intermediate designate
  %           3 different pegs
  % ensure:  print instructions to move n
  %          disks from start peg to
  %          ending peg
if (n=1) then
```

```
    write("Move disk from ", start, " peg to ",
          ending," peg")
  else
    TOH(n-1, start, intermediate, ending)
    write("Move disk from ", start, " peg to ",
          ending," peg")
    TOH(n-1, intermediate, ending, start)
  endif
  end procedure
```

Here, the parameters `start`, `ending` and `intermediate` could be integers, if the pegs are numbered, or strings, like "left," "middle," and "right." We'll use the latter in the follwoing example.

For example, if you were solving this puzzle with five disks on the left peg, you would make the initial call `TOH(5, left, right, middle)`. This is a little too complicated to trace through as a first example, so we'll try something simpler. The results with $n = 1$ are obvious, so let's trace through the call `TOH(2, left, right, middle)`. The important thing to notice is the value of the parameters in each call. The first statement executed is the call `TOH(1, left, middle, right)`, so execution in our original call is suspended momentarily, until we return from the recursive call. Now the call `TOH(1, left, middle, right)` is a base case, so `Move disk from left peg to middle peg` is printed and we return from this procedure call. Execution in our original call picks up where that call left off, and so the next statement executed is the write statement. So `Move disk from left peg to right peg` is printed. The next statement to be executed is the recursive call `TOH(1, middle, right, left)`. Again, execution in our original call is momentarily suspended until we return from this call. This call is a base case, so `Move disk from peg middle to peg right` is printed and we return from this call to our original procedure, where execution continues. But that's the end of the procedure. Thus the entire output was

```
Move disk from left peg to middle peg
Move disk from left peg to right peg
Move disk from middle peg to right peg
```

✓ Try the Tower of Hanoi with $n = 4$.

The number of moves increases rapidly, so unless you have lots of patience and time, don't start with too large of n. We will calculate exactly how many moves are made in the next chapter.

Exercises

1. Trace through the binary search when searching for 18 in the array $A[1..100]$ where $A[i] = 2i$ for $i = 1, 2, \ldots, 100$. Try it when searching for 76. For 37.
2. Write an iterative version of binary search.
3. Write a iterative version of the function `fib`. Now compare how much work is down by your version and the recursive version given in the chapter. Specifically, count the number of additions performed by your iterative version and by the recursive version when computing `fib(3)`, `fib(4)`, `fib(5)`, and `fib(6)`. Based on this very preliminary information, which version appears to be more efficient?
4. Trace through the recursive gcd function when computing `gcd(100,304)`, `gcd(101,34)`, and `gcd(1002, 248)`. How many recursive calls are made for each of these?
5. Try the recursive extended gcd: `gcdext(100,304,d,a,b)` and `gcdext(37,20,d,a,b)`.
6. Suppose $m < n$. If $\gcd(n, m) = 1$, then $m^{-1} \bmod n$ exists. Indeed, $m^{-1} \bmod n = b$, where $1 = a \cdot n + b \cdot m$ (since clearly the value of $b \cdot m = 1 \bmod n$). Thus we can find $m^{-1} \bmod n$ by using `gcdext`. Do this to calculate $11^{-1} \bmod 26$. Check your answer.
7. Calculate $10^{-1} \bmod 91$ and $5^{-1} \bmod 12$. (See the previous exercise.)

8. Use the TOH procedure to write out the solution for the four-disk TOH problem you got in the reader check. How many moves did you make? Repeat for three, two, one, and five disks. Do you have a guess for how many moves are necessary to solve the n disk problem?

9. How long will it take the Buddhist monks in heaven to solve their puzzle? (Give your answer in convenient units.) You'll need your answer from the previous exercise to calculate how many moves are necessary.

10. Suppose $A[1..n]$ is an array of integers and `target` is an integer. Write a recursive function that returns a count of how many times `target` occurs in $A[1..n]$. The header for this function should be

    ```
    function Countem(A,n,target)
    ```

    (*Hint*: The call to `Countem(A,n,target)` is to return a count of how many times `target` occurs in $A[1..n]$, so what is the simplest situation? That is, for what value of n is this easiest to do? This will be your base case. Now if it's not the base case, consider the number of occurrences of `target` in $A[1..n-1]$ (How would you find that?) then consider the value of $A[n]$ to get the count you need.)

11. Trace through this procedure. What does it do? Implement it using your favorite programming language and run it.

    ```
    procedure prob1()
      if (not the end of line) then
        read(c)                 % c is a character
        prob1()
        write(c)
      endif
    end procedure
    ```

12. What does this function do? Here A is an array of integers indexed starting at 1.

    ```
    function prob2(A, n)
      if (n = 1) then
    ```

```
      return(A[1])
    else
      return(A[n] + prob2(A, n-1))
    endif
  end function
```

13. What does this function do? Here P is usually an array of characters, although it could be an array of any type of data.

```
  function prob3(low, high, P)
    if (low > high) then
      return(true)
    else
      if (P[low] ≠ P[high]) then
        return(false)
      else
        return(prob3(low+1, high-1, P))
      endif
    endif
  end function
```

14. Suppose $A[1..n]$ is an array of integers. Write a recursive function that returns the largest integer in $A[1..n]$. The header for this function should be

    `function Max(A,n)`

    (*Hint*: Focus on the base case and the recursive call you need to make if it's not the base case.)

Programming Problems

1. Write an inductive definition for x^n for $n > 0$ and implement your definition with a recursive function.
2. Implement `gcd` in your favorite programming language. Test your program well.
3. Implement `gcdext` in your favorite programming language. Test your program well.

4. Implement `bin-search` in your favorite programming language. Test your program well.
5. Write a recursive guessing game program: The user picks a secret number between 1 and 100 and your program does the guessing. After each guess, the user responds with "too high," "too low," or "correct." Your main program should simply be one call to your recursive procedure or function.
6. Write a recursive program to implement the following definition:

$$\mathtt{B}(n,k) = \begin{cases} 1, & \text{if } k = 0 \\ 1, & \text{if } n = k \\ \mathtt{B}(n-1,k-1) + \mathtt{B}(n-1,k), & \text{if } 0 < k < n. \end{cases}$$

7. Program the Tower of Hanoi. (Be careful when you run this not to enter too large of a number or else prepare for a large number of instructions!)
8. Given positive integers a and n $(a < n)$, use `gcdext` to find $a^{-1} \bmod n$ if $\gcd(a,n) = 1$.
9. Given a simple linked list, write a recursive procedure to reverse the links in the list. The tail of the old list should be the head of the new list.

Chapter 7
Solving Recurrences

In Chapter 6 we saw several examples of recursive definitions. These mathematical formulas are called recurrences. It is easy to write a recurrence for the familiar geometric sequence. For example, consider the geometric sequence $5, 15, 45, 135, \ldots$. If a_n is the nth element of the sequence and we start indexing our numbers at 0, then we have the recurrence

$$\begin{aligned} a_0 &= 5, \text{ and} \\ a_n &= 3a_{n-1}, \text{ if } n > 0. \end{aligned}$$

We would like, if possible, to find a *closed form* for a_n; that is, a form that requires a fixed number of "simple" arithmetic computations regardless of the value of n. Finding such a form is called *solving a recurrence.* This is not an easy task in general. Indeed, not all recurrences have closed forms, but many of the recurrences we use do have closed forms.

There are many techniques, some very complex, for solving recurrences of different forms. Indeed, if you've taken a differential equations course, you know of the many methods that exist for solving differential equations. Solving recurrences has a similar large collection of techniques. We will only be concerned with one method (with many applications) in this chapter, since this is an introduction to the subject: repeated substitution (also called expansion). We'll demonstrate on the geometric sequence in the first paragraph of this chapter.

We make use of the observation that $a_n = 3a_{n-1}$ for *all* $n > 0$. In particular, if $n > 1$, then $n - 1 > 0$, and so $a_{n-1} = 3a_{n-2}$. (We get this by substituting $n = n - 1$ in the original equation.) Similarly, if $n > 2$, then $n - 2 > 0$ so $a_{n-2} = 3a_{n-3}$, and so on. The idea is to find a general pattern when making the substitutions. Assuming that $n > 0$,

$$\begin{aligned} a_n &= 3a_{n-1} = 3(3a_{n-2}) \\ &= 3^2 a_{n-2} = 3^2(3a_{n-3}) \\ &= 3^3 a_{n-3}. \end{aligned}$$

By now we see the pattern involving the exponent of 3 and the subscript of a. If we repeat this substitution k times, we see that $a_n = 3^k a_{n-k}$, a general form. Now if we let $k = n$, we have $a_n = 3^n a_{n-n} = 3^n a_0$. We picked $k = n$ because the general form then allows us to express a_n in terms of the base case a_0. We know the value of a_0 and can now substitute that value to get $a_n = 3^n a_0 = 3^n 5$, a closed form for a_n.

> ✓ Solve the recurrence where $a_0 = 1$ and $a_n = 2a_{n-1}$ for $n > 0$.

Note that this closed form is also valid for $n = 0$. Repeated substitution works well with many recurrences. Indeed, it should be your first choice when trying to solve a recurrence.

Using this method, you should be able to derive a closed form for the general form for a geometric sequence. (This is an exercise at the end of the chapter.)

As a second example, let's look at the Tower of Hanoi puzzle we wrote a solution for in the recursion chapter. We'd like to know how many moves are made when solving this puzzle with n disks. Since we wrote a recursive procedure to solve this puzzle, finding a recurrence to count the number of moves is easy. Let t_n be the number of moves to solve the Tower of Hanoi puzzle with n disks. Clearly, $t_1 = 1$. If $n > 1$, we move one disk and then solve the problem twice for $n-1$ disks. Thus, $t_n = 2t_{n-1} + 1$. Now let's solve this recurrence:

$$\begin{aligned}
t_n &= 2t_{n-1} + 1 \\
&= 2^2 t_{n-2} + 2 + 1 \\
&= 2^3 t_{n-3} + 2^2 + 2 + 1 \\
&\dots \\
&= 2^k t_{n-k} + 2^{k-1} + 2^{k-2} + \cdots + 2 + 1 \quad \text{(this is the general form)} \\
&\dots \\
&= 2^{n-1} t_1 + 2^{n-2} + 2^{n-3} + \cdots + 2 + 1 \\
&= \sum_{i=0}^{n-1} 2^i \\
&= 2^n - 1
\end{aligned}$$

(This last inequality you can prove using induction. Do so if you have not done so already.) Once we had the general form, we let $k = n - 1$, since then $t_{n-k} = t_1$, which is the base case. Thus, from this closed form we see it takes 1023 moves to solve the Tower of Hanoi puzzle with 10 disks. Of course, we've not shown our scheme is optimal. (It is.)

Frequently, recurrences arise in the analysis of an algorithm. That is, we wish to count the number of operations performed during the execution of a procedure. For example, let's take the familiar Bubble Sort:

```
procedure BubbleSort(A, n)
   % put A[1..n] in non-decreasing order
  for top=n-1 downto 1 do
    for i=1 to top do
      if (A[i]>A[i+1]) then
        swap(A[i],A[i+1])
      endif
    endfor
  endfor
end procedure
```

We wish to write a function that gives the number of statements executed when executing `Bubble Sort` on the list $A[1..n]$. Of course, this depends on exactly what elements are stored in this array. The usual position to take is to give the worst case, which is then a sort of guarantee that no matter what data are in the array, it will take no more than the number of steps we compute. Before proceeding, we must decide exactly what is meant by a statement. Certainly, there are more statements if we implement this in assembly language than in a higher-level language. We don't fret about this detail too much, arguing that the number of statements executed in one form compared to the number in another is more or less some constant multiple that depends on the form of the statements and not the algorithm. (For instance, it may be that the assembly version of an algorithm is always three times the number of C++ statements, more or less, irrespective of the algorithm.) We don't worry about this constant. You will examine this issue of analysis of algorithms in a later course.

Back to `Bubble Sort`: Let us just count the number of times the statement `swap(A[i],A[i+1])` is executed *in the worst case*. Of course, here the worst case occurs when the list is in inverted sorted order, and then this statement is executed every time through the inner loop. (We are ignoring the other statements here, partly to simplify the process, partly because the other statements are all executed the same number of times regardless of input, and partly because no other statement will be executed more than the `swap` statement in the worst case.)

Let B_n be the number of time the `swap` statement is executed in the worst case on input of size n. One iteration of the outer loop produces a list whose largest element is in $A[n]$, leaving the list $A[1..n-1]$ to be sorted on subsequent times through the loop. The second time through this outer loop also has the next-to-largest element in $A[n-1]$. In general, the kth time through the outer loop puts the k largest elements in their correct positions, $A[n], A[n-1], \ldots, A[n-k+1]$.

Note that in the worst case, the swap statement will be executed *every* time we compare $A[i]$ and $A[i+1]$. That is, the boolean statement $A[i] > A[i+1]$ will be true each time evaluated. This will happen, by the way, when the list is in decreasing order. Note that if $n = 1$, the body of the loop never executes (the list is already sorted since there is but one element) so the swap statement never executes. That is, $B_1 = 0$. Now notice that if $n > 1$, the inner loop repeats $n-1$ times on the first time through the outer loop, and so the swap statement executes $n-1$ times on the first time through the outer loop in the worst case. After this first pass, the largest element is in $A[n]$. leaving us with the list $A[1..n-1]$ yet to be sorted. This list is of length $n-1$. In other words, $B_n = n - 1 + B_{n-1}$. We now have a recurrence for B_n. Let's restate it:

$$B_1 = 0, \text{ and}$$
$$B_n = B_{n-1} + n - 1, \text{ if } n > 1.$$

Again, this recurrence holds for all $n > 1$. So, in particular, if $n > 2$, then $n - 1 > 1$ and therefore if replacing n with $n-1$ in the recurrence, we get

$$B_{n-1} = B_{n-2} + (n-1) - 1 = B_{n-2} + n - 2.$$

Similarly, if $n - 2 > 1$,

$$B_{n-2} = B_{n-3} + (n-2) - 1 = B_{n-3} + n - 3,$$

and so on.

This is the key observation to solving this recurrence using repeated substitution:

$$\begin{aligned}
B_n &= B_{n-1} + n - 1 \\
&= B_{n-2} + (n-2) + (n-1) \\
&= B_{n-3} + (n-3) + (n-2) + (n-1) \\
&\dots \\
&= B_{n-k} + (n-k) + \cdots + (n-1) \\
&\dots \\
&= B_1 + (n-(n-1)) + \cdots + (n-1) \\
&= 1 + 2 + \cdots + (n-1) \\
&= n(n-1)/2
\end{aligned}$$

Note that we got the third from last line from the general form by letting k equal $n-1$. The last equality is due to the well-known formula for the sum of the first n positive integers you proved using induction. (Here, we're summing the first $n-1$ positive integers, of course.)

As a final example, let's analyze the binary search. We call each examination of an element in the sorted list a *probe*. We want to count approximately how many probes occur in the worst case. The worst case occurs when the value we're searching for is not in the list. For a list of length one, we make one probe. If the list is of length n (larger than one), we make one probe and then recursively search a list of length $n/2$. That is, if P_n is the number of probes (in the worst case) when using the binary search to search a list of n items, then we have the recurrence $P_1 = 1$, $P_n = 1 + P_{n/2}$, for $n > 1$.

To solve this more easily, let's assume that n is a power of 2, say $n = 2^j$. Thus $j = \log n$. Then

$$\begin{aligned}
P_n &= 1 + P_{n/2} \\
&= 1 + 1 + P_{n/2^2} = 2 + P_{n/2^2} \\
&\dots
\end{aligned}$$

$$
\begin{aligned}
&= k + P_{n/2^k} \\
&\dots \\
&= j + P_{n/2^j} = j + P_1 \\
&= j + 1 = \log n + 1
\end{aligned}
$$

Thus when using the binary search to find an item in a sorted list of 1,000,000 items, we will make no more than $\log 10^6 + 1 \approx 21$ probes. Contrast that to a linear search, which would take 1,000,000 probes in the worst case. Thus the binary search is very efficient by comparison.

> ✓ If $n = 1{,}000{,}000$, approximately what is the most probes necessary when using the binary search? When $n = 1{,}000{,}000{,}000$?

Exercises

1. Solve the general recurrence for the geometric sequence:

$$
\begin{aligned}
a_0 &= d \\
a_n &= m a_{n-1}, \text{ if } n > 0.
\end{aligned}
$$

Solve these recurrences:

2. $a_1 = 2$, and $a_n = a_{n-1} + 2$, for $n > 1$.
3. $T_0 = 2$, and $T_n = T_{n-1} + 1$, for $n > 0$.
4. $a_1 = 3$, and $a_n = a_{n-1} + n$, for $n > 1$.
5. $a_0 = 4$, and $a_n = 3a_{n-1} + 1$, for $n > 0$.
6. $a_0 = 2$, and $a_n = 2a_{n-1} + 10$, for $n > 0$.
7. $a_0 = c$, and $a_n = a_{n-1} + b$, for $n > 0$, where c and b are constants.
8. $a_0 = c$, and $a_n = m a_{n-1} + b$, for $n > 1$, where c, m and b are constants.

* 9. $a_0 = 2$, and $a_n = 2a_{n-1} + n$, for $n > 0$. (You will find useful the formula in Exercise 5 from Chapter 4.)

10. $T_2 = 1$, and $T_n = T_{n-1} + 2n$, for $n > 2$.
11. $b_1 = 2$, and $b_n = b_{n/2} + 2$, for $n > 1$. (Assume that n is a power of 2 in this problem.)
12. It is a tradition in your club for a new member to shake hands with each of the current members. Suppose the club starts with one member and new members are introduced one at a time. Find a recurrence for determining the *total* number of handshakes made when the club has n members. Solve this recurrence. How many handshakes are made if membership grows to 100?
13. Suppose you put \$100 in a bank account that pays 8% interest compounded at the end of each year. Find a recurrence for the balance of your account after n years. Solve this recurrence. How much will you have in your account after 10 years? After 20 years?
14. Suppose you put \$100 in an account *every month* with an annual interest of 8% compounded every month. Find a recurrence for the balance of your account after n months. How much will your have in your account after 10 years? After 20 years?
15. Suppose your credit card company charges you 1.5% per month on any outstanding balance and you have an unpaid balance of \$1. You don't pay this piddling amount and forget about it, since you don't use this credit card anymore. Find a recurrence for how much you owe after n months. How much will you owe after 1 year? After 5 years? After 10 years?

Chapter 8
Counting

In how many ways can you arrange 10 books on a shelf? How many ways can you choose a basketball team from 8 players? How many 5-card poker hands are there? In how many ways can we line up 5 red balls and 4 blue balls? How many ways are there for interconnecting 15 computers? It is not uncommon to ask "In how many ways can we do such-and-such?" We will develop fundamental techniques to answer these types of questions.

In many cases counting is surprisingly difficult, and often the results of counts are surprisingly large. In some computer programs we might wish to generate all instances of a certain type and check some property of each instance—we call this technique a *blanket search.* Frequently, the number of instances is so large that the direct approach of examining every instance is too inefficient to be practical. Programmers are challenged to find an alternate method, one that need not check every instance but somehow takes advantage of the kind of data being processed in such a way as to use results of previously processed data when handling new cases. When programmers analyze their programs for efficiency, one aspect of their work is the counting of how many instances their algorithm must process.

Thus counting is critical to determining the efficiency of algorithms. To perform the task of counting, we'll consider a variety of techniques that take a large problem and break it into more manageable subtasks.

The Rules of Sum and Product

There are two important rules in counting that may seem obvious: the rule of sum and the rule of product.

The Rule of Sum: If the first task can be performed in m ways and the second task can be performed in n ways, and the two tasks cannot be performed simultaneously, then performing *either* task can be done in $m + n$ ways.

For example, suppose there are 5 red books and 6 blue books. Suppose the first task is to pick one red book, which can be done in 5 ways. Suppose the second task is to pick one blue book, which can be done in 6 ways. Therefore, there are $5 + 6 = 11$ ways of picking a red or blue book. (Implicit here is you can't pick a book that is both red and blue, of course.) This may be a trivial example, but be clear on when you can use this rule; not all uses will be trivial.

This rule can be extended to n tasks: If we have n tasks that can be performed and task i can be performed in m_i ways and no two tasks can be performed simultaneously, then performing any one of the tasks can be done in $m_1 + m_2 + \cdots + m_n$.

Here's another situation where rule of sum is used: Suppose we wish to find how many subsets of six books there are, where each subset has at least four books. The task of picking a subset of at least four books can be done in three different ways: We could pick a subset of exactly four books, or pick a subset of exactly five books, or pick a subset of exactly six books. Note these can't be performed simultaneously. Let B_n, for $n = 4$, 5, or 6, be the number of subsets of size n. Therefore, the count we're interested in is $B_4 + B_5 + B_6$. (We will find what the value of each of these is later.)

When programmers analyze sequential code, they use this rule. For each task in the sequence of steps, the programmer associates some performance expression, say execution time, and then adds those expressions to get the total performance expressions for that code segment.

The Rule of Product: If a task can be broken into a first and a second stage and the first stage can be performed in m ways and the second stage can be performed in n, then the task can be performed in mn ways.

For example, suppose there are 6 men up for the male lead of a play and 8 women up for the female lead. We want to know how many ways there are to cast the two leads. We break this task into two stages: First cast the male, then cast the female. There are 6 ways to do the first and 8 ways to do the second, making $6 \cdot 8 = 48$ ways to cast the two leads. (We could have first cast the female then the male, of course; the result would be the same.)

This rule can also be generalized if a task can be decomposed into more than two stages. If the task can be broken into n stages and task i can be performed in m_i ways, then the task at hand can be done in $m_1 m_2 \cdots m_n$. For example, suppose a license plate is made up of 2 *different* letters followed by 3 *different* digits. We break the task of making a license plate into 5 stages: First pick the first letter, then pick the next letter, then the first number, then the second, and finally the third. The rule of product says this can be done in $26 \cdot 25 \cdot 10 \cdot 9 \cdot 8$ =468,000 ways. If we don't insist that the letters and numbers be different, then the number of license plates would be $26 \cdot 26 \cdot 10 \cdot 10 \cdot 10 = 676{,}000$.

✓ In your closet you have eight shirts, four pants, and three belts. In how many ways can you outfit yourself if an outfit consists of a shirt, pants, and a belt?

Permutations

Suppose we have n distinct objects. That is, we can distinguish between any two of the n objects. How many ways can we arrange these n objects in a row? We can break this task into stages: Pick the first object, then the second, then the third, and so on. Using the rule of product, we see that there are $n \cdot (n-1) \cdots 2 \cdot 1 = n!$ ways of arranging these n objects. Such an arrangement is called a *permutation.* Thus, there are $10! = 3{,}628{,}800$ ways of arranging 10 books on a shelf. (This might be surprisingly large to you.)

Now suppose we pick m objects from these n ($m \leq n$) and arrange them. How many ways can this be done? Proceeding as before, we see there are $n \cdot (n-1) \cdots (n-m+1)$ ways to do this. Note that this equals $n!/(n-m)!$. We call this the number of permutations of m objects taken from n and denote it by $P(n, m)$.

> ✓ You have 10 shirts and are going to pack 3 shirts in a suitcase for a trip. You'll be packing so the shirt for the third day is on bottom, the shirt for the second day is in the middle, and the shirt for the first day is on top. In how many ways can you stack 3 shirts for your trip?

How many permutations are there of the letters in `COMPUTER`? Since each letter is distinguishable, the preceding argument shows us that there are 8! permutations. How many permutations are there of the letters `BALL`? This is different since the two `L`'s are not distinguishable. To count this, let's first suppose the two `L`'s are distinguishable, count the permutations, then adjust the count. So, let's label one $\mathtt{L}_1$ and the other $\mathtt{L}_2$. Then there are 4! different permutations of the four distinguishable letters. However, now regarding the two `L`'s as indistinguishable, we see that the 4! has counted some permutations more than once. For instance, $\mathtt{L}_1\mathtt{ABL}_2$ is indistinguishable from $\mathtt{L}_2\mathtt{ABL}_1$, but we have counted it twice. Likewise, we have counted every indistinguishable permutation twice. Hence the total number of indistinguishable permutations of the letters `BALL` is $4!/2$.

Similarly, consider the total number of distinguishable permutations of the letters `RECEIVE`. If we count the three `E`'s as distinguishable, then there are 7! permutations. But notice that each permutation of *indistinguishable* `E`'s has been counted 3! times, since the three `E`'s can appear in any one of 3! orders. Thus there are $7!/3!$ different permutations of these letters, that are distinguishable.

To generalize further, consider the letters in `PEPPER`. There are two `E`'s, three `P`'s, and an `R`. First thinking of all six letters as distinguishable, then dividing by the number of times we've counted each permutation, where the same letters appear but in different orders, we see we have a total of $6!/(2!\,3!)$ distinguishable permutations.

In general, if $n = n_1+n_2+\ldots+n_k$ and we have n objects of k types and any two items of the same type are indistinguishable and there are n_i objects of the ith type, for $i = 1, \ldots, k$, then there are $n!/(n_1!\,n_2!\ldots n_k!)$ distinguishable permutations of the n objects. For example, there are $11!/(1!\,4!\,4!\,2!)$ (distinguishable) permutations of the letters in `MISSISSIPPI`. This formula has a surprising number of applications in counting.

Consider the problems posed at the beginning of the chapter asking how many ways we can line up 5 red balls and 4 blue balls. This is equivalent to the problem of how many words can be made from 5 R's and 4 B's. We now know this is $9!/(5!4!) = 126$.

> ✓ How many distinguishable rearrangements are there of the letters in `BUTTERBALL`?

Combinations

Suppose we have a standard deck of 52 cards and deal a 5-card hand. How many 5-card hands are there? Note that we are interested in counting the number of hands, not the number of ways we can deal 5 cards. The difference is that the latter situation depends on the order the cards are dealt, while in the former we are not concerned with the order the cards were dealt to us, only what cards we end up with in our hand. Keep in mind that the number of ways of dealing 5 cards is $P(52, 5) = 52 \cdot 51 \cdot 50 \cdot 49 \cdot 48$.

Every 5-card hand can be dealt in 5! different ways. Thus,

$$5! \times \text{the number of 5-card hands}$$

$$= \text{the number of ways of dealing 5 cards}$$
$$= 52 \cdot 51 \cdot 50 \cdot 49 \cdot 48.$$

So the number of 5-card hands is $(52 \cdot 51 \cdot 50 \cdot 49 \cdot 48)/5!$. Note that this is equal to $52!/(47!\,5!)$.

Another way to think of this problem is to count the number of subsets of size 5 from a set of 52 objects. (We call subsets of size 5, 5-subsets.) We call this count "52 choose 5" or "combinations of 52 things taken 5 at a time" and denote it $\binom{52}{5}$ or $C(52, 5)$.

In general, the number of k-subsets we can get from a set of n items ($k \leq n$, of course) is

$$C(n,k) = \binom{n}{k} = \frac{n!}{(n-k)!\,k!}.$$

Note that $\binom{n}{0} = 1$. (Recall that $0! = 1$.) Also, $\binom{n}{1} = \binom{n}{n} = 1$.

✓ Verify that $\binom{n}{0} = 1$, $\binom{n}{1} = 1$ and $\binom{n}{n} = 1$.

Notice that $\binom{n}{k} = \binom{n}{n-k}$ since

$$\binom{n}{n-k} = \frac{n!}{(n-(n-k))!\,(n-k)!} = \frac{n!}{k!\,(n-k)!} = \binom{n}{k}.$$

This makes sense since whenever we pick k objects from n, we are not picking the remaining $n-k$ objects. If we think of picking objects as putting them in basket A and putting the remaining objects in basket B, then every k-subset in basket A corresponds to a $(n-k)$-subset in basket B. Thus these two counts should be equal.

To avoid confusion between permutations and combinations remember that with permutations order matters. With combinations order does not matter. Here are some more examples to illustrate the use of combinations:

1. How many arrangements of the letters in `TALLAHASSEE` are there that have no adjacent `A`'s? We'll count this by first arranging the other eight letters, then inserting the three `A`'s in the nine available spots. Note that there are nine positions to put the `A`'s after the other eight letters have been decided on: at the beginning, between the first and second letters, between the second and third, and so on. This will guarantee that there are no adjacent `A`'s. Note that the rule of product applies here. As we have seen before, there are $8!/(2!\,2!\,2!)$ ways of accomplishing the first task. Now we simply must choose three of the available nine positions and there are $C(9,3) = 9!/(6!\,3!)$ ways of doing this for a total of

$$\frac{8!}{2!\,2!\,2!} \cdot \frac{9!}{6!\,3!} = 423{,}360.$$

2. Suppose there are 53 students and you must pick a nine-person volleyball team. How many ways can this team be picked? This is just $C(53,9) = 53!/(44!\,9!) = 4{,}431{,}613{,}550$. Now suppose we have 36 people that we must split into 4 volleyball teams. How many ways can we do this? We can first pick one team from the 36 people, then pick the next team from the remaining 27, then the third team from the remaining 18, and finally the last team from the remaining 9. By the rule of product, this is

$$\begin{aligned}\binom{36}{9}\binom{27}{9}\binom{18}{9}\binom{9}{9} &= 94{,}143{,}280 \cdot 4{,}686{,}825 \cdot 48{,}620 \cdot 1 \\ &= 21{,}452{,}752{,}266{,}265{,}320{,}000.\end{aligned}$$

Another way to count here is to think of naming the four teams `A`, `B`, `C`, and `D`. The preceding solution assigns people to teams. Instead we could assign teams to people. Think of lining up the 36 people and assigning each a letter. This is just counting the number of arrangements of four `A`'s, four

`B`'s, four `C`'s, and four `D`'s. This is $36!/(9!\,9!\,9!\,9!)$, which is the same number we calculated using the other method, as you can verify.

3. Suppose there are 6 boys and 5 girls and you wish to choose a basketball team that has either 2 or 3 boys on the team. How many such teams are there? Here is an example where the rule of sum applies. We will count how many teams with 2 boys, how many teams with 3 boys, and then, by rule of sum, add these counts to get how many teams with 2 or 3 boys. To pick a team with 2 boys, we first pick the 2 boys, then pick the 3 girls. By the rule of product this is

$$C(6,2) \cdot C(5,3) = \frac{6!}{4!\,2!}\frac{5!}{2!\,3!} = 150.$$

Similarly, the number of teams with 3 boys and 2 girls is 200, for a total of 350 different teams.

> ✓ Verify that the number of teams with 3 boys and 2 girls is 200.

4. How many 5-card poker hands are there with exactly one pair and no higher? We will count this in the following way: First we'll pick the denomination for the pair, then pick the two cards from the four in that denomination, then pick the remaining three cards. The difficulty comes in picking the remaining three cards. We must make sure we don't pick a card that is the same as our pair, and we must not pair up any of the three other cards. But we can count this by picking the third card from the 48 remaining cards that don't match the first two cards, then picking the fourth card from the remaining 44 cards that match none of the first three cards picked, and finally picking the fifth card from the remaining 40 cards that don't match the first four cards. This would give $48 \cdot 44 \cdot 40$ ways of picking the remaining 3 cards. But hold on—we've counted permutations, not combinations, of these 3 cards here. That is, each legal 3 card set has been counted $3!$ times. Thus there

are $(48 \cdot 44 \cdot 40)/3!$ legal 3-card hands. Putting this all together (using the rule of product, of course), we get

$$13 \cdot \binom{4}{2} \cdot \frac{48 \cdot 44 \cdot 40}{3!} = 1{,}098{,}240$$

different 5-card poker hands with exactly one pair.

> ✓ How many 5-card poker hands are there with exactly one three-of-a-kind and no higher?

Calculation Considerations

When calculating $\binom{n}{k}$, the most obvious approach is to compute $n!$, $(n-k)!$, and $k!$ and then perform the necessary arithmetic. This is inefficient, and it may be that $n!$ will cause integer overflow while the final answer is relatively small. To overcome this, observe that

$$\binom{n}{k} = \frac{n!}{(n-k)!\,k!} = \frac{n(n-1)(n-2)\cdots(n-k+1)}{k(k-1)\cdots 2 \cdot 1}$$
$$= \frac{n(n-1)(n-2)\cdots(n-k+1)}{1 \cdot 2 \cdots (k-1)k}.$$

The form of the last expression is where we take our inspiration for an efficient method of calculation. We start calculating our total by initially assigning n as our temporary value, then dividing by 1, then multiplying by $n-1$, then dividing by 2, then multiplying by $n-2$, then dividing by 3, and so on. Here's the algorithm:

```
function Comb(n, k)
 combo = 1;
 for i=1 to k do
   combo = combo*(n+1-i);
   combo = combo/i;
 endfor;
 return(combo);
end function
```

There are two things to note: (1) The partial answers are never much larger than the final answer, and (2) each partial calculation is an integer. This later point is not immediately obvious. But consider a couple of partial calculations. After cycling through the for loop twice, we've calculated

$$\frac{n(n-1)}{1 \cdot 2}.$$

Note that $n(n-1)$ is even, so when we divide by 2, the result is an integer. Likewise,

$$\frac{n(n-1)(n-2)}{1 \cdot 2 \cdot 3}$$

is an integer since $n(n-1)(n-2)$ is divisible by 6. Why? Because at least one of the consecutive integers n, $n-1$, and $n-2$ is even and at least one of the these integers is divisible by 3 (since these are three *consecutive* integers). This idea carries through to all the intermediate calculations.

✓ Calculate $\binom{10}{4}$ using the function `Comb`.

The Binomial Theorem

Another important aid in counting is the Binomial Theorem. You may recall from algebra that

$$\begin{aligned}(x+y)^n &= \binom{n}{0}y^n + \binom{n}{1}xy^{n-1} + \binom{n}{2}x^2y^{n-2} + \cdots \\ &\qquad + \binom{n}{n-1}x^{n-1}y + \binom{n}{n}x^n \\ &= \sum_{k=0}^{n}\binom{n}{k}x^k y^{n-k}.\end{aligned}$$

For example, $(x+y)^3 = y^3 + 3xy^2 + 3x^2y + x^3$. Since the coefficients used in the expansion are $\binom{n}{k}$, these are frequently called the *binomial coefficients.* As we've seen before, these also count the number of combinations of n things taken k at a time.

An important consequence of the Binomial Theorem is that $\sum_{k=0}^{n} \binom{n}{k} = 2^n$. Why? Just let $x = y = 1$ in the Binomial Theorem. For instance,

$$\binom{4}{0} + \binom{4}{1} + \binom{4}{2} + \binom{4}{3} + \binom{4}{4} = 2^4 = 16$$

as you can easily check directly.

✓ Expand $(x+y)^4$ using the Binomial Theorem.

Applications of Counting to Probability

Suppose you draw a card from a shuffled standard deck of 52. You'd like to have some measurement of how likely it is for that card to be a face card. We say that the act of drawing the card from the 52-card deck is an *experiment* and the set of all possible outcomes is the *sample space.* Here, the sample space is {AS, KS, QS, JS, TS, 9S, ..., AC, KC, ..., 3D, 2D}, which has 52 items. In this experiment, it is reasonable to assume that each possible outcome is equally likely. There are certainly many experiments where this is not the case. An example of such an experiment is to draw the card and then record its denomination, with the denomination of an ace being 1 and the denominations of all face cards being 10. Here, the sample space is $\{1, 2, 3, 4, 5, 6, 7, 8, 9, 10\}$, but certainly the outcome of 10 is much more likely than any of the other outcomes. We will limit ourselves to those cases where all outcomes are equally likely.

An *event* is simply a subset of the sample space. In our example, the event is the set of all face cards. Suppose S is the sample space and E is an event. (That is, $E \subseteq S$.) If each outcome in S is equally likely, then the

$$\text{probability that } E \text{ occurs} = \Pr(E) = \frac{|E|}{|S|}.$$

Thus to compute $\Pr(E)$, we need to count how many elements there are in S and in E. In our example, $|S| = 52$ and E is the set of all face cards, so $|E| = 12$ and thus $\Pr(E) = 12/52 = 3/13 \approx 0.231$. $\Pr(E)$ indicates what percentage of time the experiment will yield an outcome in the event E *in the long run.* In our example, we'd expect that about 23% of the time we would draw a face card in the long run.

✓ What's the probability of drawing an even numbered card from the deck?

The probability of any event has to lie between 0 and 1, inclusive. We call an event E where $\Pr(E) = 0$ an *impossibility* and an event E where $\Pr(E) = 1$ a *certainty.* If $\Pr(E) > 1/2$, then E is more likely than not to occur; we sometimes say E has a better than even chance of occurring. If $\Pr(E) < 1/2$, then E is more likely not to occur. From our previous example, we see that it is more likely not to draw a face card than draw one, since $\Pr(\text{drawing a face card}) = 3/13 < 1/2$.

The *complement* of an event E, $\overline{E}$, is all the outcomes not in E. That is, $\overline{E} = S - E$. Clearly, $\Pr(\overline{E}) = 1 - \Pr(E)$. Thus the probability of not drawing a face card in our experiment is $1 - 3/13 = 10/13$.

We'll conclude this chapter by computing some probabilities.

What's the probability of being dealt a 5-card poker hand with exactly one pair and no higher? Here, S is the set of all 5-card hands and E is the set of 5-card hands that contain exactly one pair and no more. We've seen previously that $|S| = \binom{52}{5} = 2{,}598{,}960$ and $|E| = 1{,}098{,}240$. Thus, the probability of being dealt a hand with one pair and no more is

$$\frac{|E|}{|S|} = \frac{1{,}098{,}240}{2{,}598{,}960} \approx 0.4226,$$

which means that in the long run, you should expect to get exactly one pair and no higher about 42% of the time.

If we have 6 boys and 5 girls and choose 5 students from this group at random to form a basketball team, what's the probability that the team will have either 2 or 3 boys? Here, E is the set of all possible basketball teams formed with either 2 or 3 boys on them. As we have seen before, $|E| = 350$. The sample space, S, is the set of all possible teams formed from the 11 students. Thus $S = \binom{11}{5} = 462$ an so the probability that the team will have either 2 or 3 boys is $350/462 \approx 0.758$.

What's the probability that a 3-digit number picked at random has distinct digits? The 3-digit numbers run from 100 through 999. Thus $|S| = 900$. E is the set of 3-digit numbers with distinct digits. We count these by first picking the most significant digit, then the next digit, and finally the least significant digit. There are 9 choices for the most significant digit (0 can't be the most significant digit). Once that digit has been picked, there are 9 possible digits to pick for the next digit that are different from the first digit. Once these two digits have been picked, there are 8 possible digits to pick for the last digit that differ from the first two digits. Thus, by the rule of product, $|E| = 9 \cdot 9 \cdot 8 = 648$. So the probability that a 3-digit number picked at random has distinct digits is $648/900 = 0.72$.

Exercises

1. You go to the magazine table at your dentist's office and gather 5 magazines from the 12 on the table (all at least one year old). How many different collections of magazines can you gather?

2. Braille symbols are made by raising up from one to six of the dots in a two-by-three grid $\begin{smallmatrix}\bullet & \bullet \\ \bullet & \bullet \\ \bullet & \bullet\end{smallmatrix}$. How many different Braille symbols are there? How many symbols with at least three raised dots? How many symbols with an even number of raised dots?
3. A committee of 12 is picked from 10 men and 10 women. How many committees are there? How many if there are required to be 6 men and 6 women on the committee? How many if there is an even number of women? How many if there are more women than men? How many if there are at least eight men?
4. If a committee in the previous exercise is picked at random, what is the probability that the committee would have more women than men on it?
5. How many 5-card poker hands are there that are a flush (all five cards of the same suit)? What's the probability that a you would be dealt a flush?
6. How many 5-card poker hands are there that are 4-of-a-kind? What's the probability that you would be dealt 4-of-a-kind?
7. How many 5-card poker hands are there that have three aces and two jacks?
8. How many 5-card poker hands are there that are full houses (one pair and another 3-of-a-kind)? What's the probability that you would be dealt a full house?
9. How many ways can you get two pairs but nothing more? (The pairs must be different denominations.) What's the probability that you would be dealt two pair but nothing more?
10. You have 15 books. How many ways are there to arrange them on two shelves so that each shelf must have at least one book? If each shelf must have at least two books?

11. A computer operator has 12 rather long programs to run on a machine that does not time share. The operator must decide in which order to process the programs. In how many ways can these programs be ordered if (a) there are no restrictions? (b) four of the programs are high priority and must be done before the others? (c) the jobs are separated into 4 high priority, 5 middle priority, and 3 low priority jobs?
12. Suppose we have 6 boys and 5 girls and we randomly select a basketball team from among them. What's the probability that the team will have more girls than boys? What's the probability the team will have more boys than girls? Are these two events complementary?
13. Which has more arrangements for all the letters in its name, PENNSYLVANIA or MASSACHUSETTS?
14. How many different paths in the xy plane are there from $(0, 0)$ to $(7, 7)$ if a path proceeds one step going either right or up one unit?
15. How many integers can be formed using the digits 3, 4, 4, 5, 5, 6, 7 if we want the number to exceed 5,000,000?
16. How many 6-digit integers are there? How many if no digit is repeated? Answer these questions if the number is to be divisible by 5.
17. Assuming that a phone number has 10 digits in it, find the probability of getting a phone number with a 3 in it, assuming no restrictions on what digits are allowed.
18. How many two-digit numbers contain the digit 5? How many such three-digit numbers? How many such four-digit numbers? How many such n-digit numbers?
19. What's the probability of an n-digit number containing a digit 5? What does this probability approach as n approaches infinity? (Explain why your answer makes sense here.)
20. Twenty slips of paper numbered 1 through 20 are placed

in a hat and two are drawn out. How many different pairs of numbers can we draw out? How many of these are adjacent numbers? What's the probability that the two numbers will be adjacent?

21. A bag has 24 balls—6 each of orange, white, red, and yellow. A juggler randomly picks three balls to juggle. How many ways can this be done? How many ways where all three balls are the same color? What's the probability that all three balls are the same color?
22. Using the same bag of balls as in Exercise 21, how many ways are there to pick three balls where all three balls are different colors? What's the probability that all three balls are different colors?
23. How many distinct arrangements of the letters in *mathematics* are there?
24. How many distinct license plates can be made if each should have 3 digits followed by 3 letters?
25. How many distinct phone numbers (seven digits) can be made if the first digit is not allowed to be one and all other digits can be any of the values from zero to nine?
26. Gertrude bought six different CDs to give as gifts. How many different ways can she distribute the CDs so that each of her three boyfriends will receive two CDs?
27. A gallery owner plans to exhibit seven pieces of art in one area of her gallery, three impressionist paintings, three cubist paintings, and one realistic portrait. How many different arrangements of types paintings can she make?
28. How many numbers greater than 50,000,000 can be formed by rearranging the digits of the number 13,979,397?
29. There are three pennies, three nickels, and three dimes on a table. A child picks three coins at random. What's the probability the child picks three different denomination of coin?

30. In planning the breakfast menus for the coming week, Don notes that there are four different kinds of cereal in the pantry. How many different choices for the seven-day period can be made?

31. A box contains 20 crayons, no two of the same color. In how many different ways can the crayons be given to four children so that each gets five crayons?

32. Find the probability of getting a number divisible by 3 when a die is rolled.

33. If a coin is tossed four times, find the probability that it will land on heads all four times.

34. If a coin is tossed five times, find the probability that it will land on tails exactly three times.

35. If four coins are tossed, find the probability that all of them land with the same side up.

36. Find the probability that a phone number (seven digits, with all digits equally likely to appear) contains exactly two 5's.

37. Suppose three people are chosen at random from a group of five men and six women. Find the probability that all three are men.

38. In a consumer preference test, 10 people are asked to choose their favorite fruit from among apples, bananas, and oranges. What is the probability that nobody chooses bananas? (Assume that each fruit is equally likely to be chosen.)

39. What is the probability that a randomly chosen four-digit number contains no repeated digits?

40. Suppose two dice are thrown. Find the probability that both land with the same number of dots facing up.

41. When two dice are thrown, find the probability that the sum of the dots is equal to 9.

42. When two dice are thrown, find the probability that the sum of the dots is greater than or equal to 6.
43. What's the coefficient of $x^{80}y^{20}$ in the expansion of $(x + y)^{100}$?
44. How many eight-letter words (not necessary real words), with no repeated letters, are there if the words have three, four, or five vowels? (Assume the vowels are `A`, `E`, `I`, `O`, and `U`.)
45. How many ways can you pick three books to buy from a group of five?
46. Suppose you pick three books to buy from a group of five and $10 has been secretly slipped into one of the books. What's the probability you'll end up with that book given that you select your three books at random?
47. How many ways can you divide eight blood samples into two groups of four samples each? How many ways if each group of the two groups must have at least one sample?
48. How many odd integers between 1000 and 9999 have distinct digits?
49. A fair coin is flipped 10 times.
    a. How many sequences of heads and tails are there? (Note that each sequence is equally likely, if the coin is fair.)
    b. How many different sequences have exactly three heads? (Think of picking where the heads appear in the sequence.)
    c. What's the probability that a given sequence has exactly three heads?
    d. How many sequences have no more than three heads?
    e. What's the probability that a given sequence has no more than three heads?
    f. How many sequences have exactly five heads?
    g. What's the probability that a given sequence has exactly five heads?

Programming Problems

1. Implement the function `Comb` in your favorite programming language.
2. Write a program that inputs a word and outputs a count of how many distinguishable permutations of the letters in the word there are.
3. Write a function that simulates the flipping of a coin. Design a program that checks to see if your function seems to be simulating a good coin (i.e., one that returns heads roughly half the time and tails half the time).
4. Write a function that simulates the tossing of a die and test to see if your function represents a fair die.

Chapter 9
Matrices

Rectangular arrays of numbers show up virtually everywhere in mathematics. We call these arrays matrices. For example,

$$A = \begin{pmatrix} 1 & 2 & 3 \\ 4 & 5 & 6 \\ 7 & 8 & 9 \end{pmatrix}$$

is a three-by-three matrix with integer entries. The three horizontal collections of entries are called *rows*, while the vertical ones are called *columns*. This particular example is a square matrix, but it is not necessary that a matrix have the same number of rows as columns. If a matrix has m rows and n columns, we say it has dimension m *by* n, written $m \times n$. The following is notation often used to represent a general matrix of m rows and n columns:

$$A = \begin{pmatrix} a_{11} & a_{12} & \dots & a_{1n} \\ a_{21} & a_{22} & \dots & a_{2n} \\ \vdots & \vdots & \ddots & \vdots \\ a_{m1} & a_{m2} & \dots & a_{mn} \end{pmatrix}.$$

A matrix with only one row or column is called a *vector*. For example, $(1 \quad 3 \quad 0 \quad -1)$ is a (row) vector of dimension 4, sometimes called a 4-vector. The notation for a general (row) vector of dimension n is $(a_1\, a_2\, \dots\, a_n)$.

The entries, a_{ij}, in all of our applications will come from numbers we are already familiar with, such as integers, reals, and integers modulo n. For example, the first example we gave of a matrix has entries from the integers. So we'll say this is a 3×3 matrix over $\mathbb{Z}$. In general, if the entries come from a system S, we'll talk of the $m \times n$ matrices over S. It is important to keep in mind the set from which the entries have been taken when performing operations on matrices. In performing operations on matrices, it is necessary to use operations defined for the sets over which the matrices are defined.

Matrix Operations

There are several operations on matrices, each depending on operations of the underlying sets in some way. We can combine two matrices by adding or multiplying them, provided their dimensions are compatible. We add two matrices as follows:

$$\begin{pmatrix} a_{11} & a_{12} & \dots & a_{1n} \\ a_{21} & a_{22} & \dots & a_{2n} \\ \vdots & \vdots & \ddots & \vdots \\ a_{m1} & a_{m2} & \dots & a_{mn} \end{pmatrix} + \begin{pmatrix} b_{11} & b_{12} & \dots & b_{1n} \\ b_{21} & b_{22} & \dots & b_{2n} \\ \vdots & \vdots & \ddots & \vdots \\ b_{m1} & b_{m2} & \dots & b_{mn} \end{pmatrix} =$$

$$\begin{pmatrix} a_{11}+b_{11} & a_{12}+b_{12} & \dots & a_{1n}+b_{1n} \\ a_{21}+b_{21} & a_{22}+b_{22} & \dots & a_{2n}+b_{2n} \\ \vdots & \vdots & \ddots & \vdots \\ a_{m1}+b_{m1} & a_{m2}+b_{m2} & \dots & a_{mn}+b_{mn} \end{pmatrix}.$$

This addition is probably just what you would expect. Entries in the result matrix are obtained entry by entry using the scalar addition of the underlying set. By now you are not surprised at the use of the "+" symbol in two different ways. The "+" between the matrices indicates matrix addition, while the "+" inside the matrices indicates addition of the entry elements. We have seen overloading of symbols before and recognize that the distinction is important. One other note of importance is that when two matrices are added, both must have the same dimensions; otherwise addition is undefined.

For example,

$$\text{if} \quad A = \begin{pmatrix} 3 & 5 & 7 & 9 \\ 2 & 4 & 3 & 2 \\ 6 & 1 & 0 & 1 \end{pmatrix} \quad \text{and} \quad B = \begin{pmatrix} 2 & 1 & 4 & 0 \\ 3 & 0 & 1 & 5 \\ 1 & 2 & 0 & 3 \end{pmatrix},$$

$$\text{then} \qquad A+B = \begin{pmatrix} 5 & 6 & 11 & 9 \\ 5 & 4 & 4 & 7 \\ 7 & 3 & 0 & 4 \end{pmatrix}.$$

✓ If $A = \begin{pmatrix} 2 & 5 & 8 & 2 \\ 7 & 4 & 8 & 1 \end{pmatrix}$ and $B = \begin{pmatrix} 3 & 6 & 1 & 0 \\ 0 & 4 & 9 & 2 \end{pmatrix}$, find $A+B$.

When studying new operations of any system, we usually ask about identities and inverses. Both questions make sense with respect to matrix addition. For any collection of $m \times n$ matrices there is an additive identity consisting of the $m \times n$ matrix whose entries are each the additive identity of the entry set. (The additive identity of the integers, reals, and integers modulo n is 0, of course.) For example, if we are working with 3×4 matrices over $\mathbb{Z}_{17}$, then the additive identity is the 3×4 matrix all of whose entries are the zero element in $\mathbb{Z}_{17}$. We call this the 3×4 zero matrix, naturally enough.

Every matrix has an additive inverse. If A is an $m \times n$ matrix, then the additive inverse of A, written $-A$, is the $m \times n$ matrix formed so that $A + (-A)$ is the $m \times n$ zero matrix. It is easy to see that if a_{ij} is an entry in A, then the corresponding entry in $-A$ is $-a_{ij}$. For example, suppose we are considering 2×3 matrices over $\mathbb{Z}_3$.

$$\text{If} \quad A = \begin{pmatrix} 1 & 2 & 0 \\ 2 & 1 & 1 \end{pmatrix} \quad \text{then} \quad -A = \begin{pmatrix} 2 & 1 & 0 \\ 1 & 2 & 2 \end{pmatrix},$$

$$\text{since} \quad A + (-A) = \begin{pmatrix} 0 & 0 & 0 \\ 0 & 0 & 0 \end{pmatrix}.$$

The addition of the entries is done in $\mathbb{Z}_3$. It is of interest to note that we have used the same symbol for adding integers, "+", as we used for adding elements of $\mathbb{Z}_3$. In fact, we used the same symbol to indicate the addition of matrices. This is not surprising, nor is it unusual. It is common practice to use a particular symbol, such as "+", to indicate an operation in some mathematical system. To be precise, we would need to subscript the symbol or use a different symbol to clarify which system we are applying it to. For example, when adding in $\mathbb{Z}_3$, we could use $+_{\mathbb{Z}_3}$. While this is an excellent way to clarify where the addition is taking place, readers usually know from the context what is meant. We can just say that x and y are members of $\mathbb{Z}_3$ and then use the expression $x + y$ without causing confusion. On the other hand, if we were in the midst of a discussion involving both $\mathbb{Z}_3$ and $\mathbb{Z}_5$ and we wanted to write the expression $1 + 2$, it would be necessary to make sure that readers would know in which system the addition was to take place.

Similarly, in writing computer programs, we often apply this practice of using the same symbol to mean different things depending on how its used. We call this practice "overloading." It's a good word to express the idea that we have loaded a particular symbol with more than one meaning. For example, in C++, if we write `int x,y;` `float r,s;` and then later want to perform addition, we can write `x = x + y;` `r = r + s;` and get proper sums to be found, even though we are performing integer addition in the first situation, but real number addition in the second.

To make programs more user friendly and more readable, programmers can overload operators for other types as well. For example, if we want to write a program that adds matrices, we can overload the operator "+" so that if A and B are declared as matrices, we can write "$A + B$" to indicate that matrix addition is to be performed. We will not put in the syntax for that here, but you can easily find it in any programming language manual.

✓ Find the additive inverse of

$$A = \begin{pmatrix} 2 & 5 & 6 \\ 3 & 4 & 2 \end{pmatrix},$$

where the entries are from $\mathbb{Z}_7$.

Although there are many applications using matrix addition, matrix multiplication is a binary operation with more interesting applications. If A and B are two matrices, their product is defined only under the condition that the number of columns of A is the same as the number of rows of B. Hence, any square matrices with the same number of rows and columns can be multiplied together.

To define $AB = C$, where A is an $m \times n$ matrix and B is an $n \times p$ matrix, we note that C will be an $m \times p$ matrix with each entry defined as follows:

$$c_{ik} = \sum_{j=1}^{n} a_{ij} b_{jk}.$$

For example, suppose

$$A = \begin{pmatrix} 1 & 2 \\ 3 & 4 \\ 5 & 6 \end{pmatrix} \quad \text{and} \quad B = \begin{pmatrix} 4 & 3 & 2 & 1 \\ 8 & 7 & 6 & 5 \end{pmatrix}.$$

Since A has three rows and two columns and B has two rows and four columns, the number of columns of A is the same as the number of rows and B, and so the matrix product AB is defined and will have three rows and four columns. Using the definition given, we obtain the product AB, a matrix with three rows and four columns:

$$\begin{pmatrix} 1 \cdot 4 + 2 \cdot 8 & 1 \cdot 3 + 2 \cdot 7 & 1 \cdot 2 + 2 \cdot 6 & 1 \cdot 1 + 2 \cdot 5 \\ 3 \cdot 4 + 4 \cdot 8 & 3 \cdot 3 + 4 \cdot 7 & 3 \cdot 2 + 4 \cdot 6 & 3 \cdot 1 + 4 \cdot 5 \\ 5 \cdot 4 + 6 \cdot 8 & 5 \cdot 3 + 6 \cdot 7 & 5 \cdot 2 + 6 \cdot 6 & 5 \cdot 1 + 6 \cdot 5 \end{pmatrix}$$

$$= \begin{pmatrix} 20 & 17 & 14 & 11 \\ 44 & 37 & 30 & 23 \\ 68 & 57 & 46 & 35 \end{pmatrix}.$$

✓ Find the products AB and BA, where

$$A = \begin{pmatrix} 1 & 2 & 3 \\ 3 & 4 & 5 \\ 5 & 6 & 7 \end{pmatrix} \quad \text{and} \quad B = \begin{pmatrix} 4 & -3 & 0 \\ 0 & -2 & 6 \\ 0 & 1 & -1 \end{pmatrix}.$$

As you have just seen from the preceding example, matrix multiplication is not commutative. It's even worse than that. For example, if A is a 2×3 matrix and B is 3×4, AB is defined, but BA is not even defined.

Finding a multiplicative identity for square matrices is not difficult, although not so obvious as the additive identity. Consider the matrix

$$I = \begin{pmatrix} 1 & 0 & \dots & 0 \\ 0 & 1 & \dots & 0 \\ \vdots & \vdots & \ddots & \vdots \\ 0 & 0 & \dots & 1 \end{pmatrix}.$$

I is called the *identity matrix.* As an exercise you will show that for a given square matrix, a matrix of the form given by I serves as a multiplicative identity; that is, $AI = IA = A$ for every matrix A with the same dimension as I. So matrix I commutes with any square matrix of the same dimensions. For matrices having a different number of rows from the number of columns, multiplicative identities do not exist. This is because multiplicative identities are required to commute with all elements (in this case, matrices) in the system. A technicality perhaps, but for this reason we will only be speaking of square identity matrices.

The problem of finding a multiplicative inverse for a square matrix is considerably more difficult than finding an additive inverse. Indeed, some square matrices do not even have multiplicative inverses! We'll ask you to investigate this in the Exercises. Before finding a way to determine if a matrix has an inverse or not and how to find the inverse, if it does have one, we need to learn how to solve a system of equations.

Before doing so, we define one more operation that involves a matrix and a single element from the underlying system S. In this context, a single element from S is called a *scalar.* This operation is called *scalar multiplication* and is defined as follows. If c is an element from S (and assuming multiplication is defined on S), then cA is defined by

$$\begin{aligned} cA = c \begin{pmatrix} a_{11} & a_{12} & \dots & a_{1n} \\ a_{21} & a_{22} & \dots & a_{2n} \\ \vdots & \vdots & \ddots & \vdots \\ a_{m1} & a_{m2} & \dots & a_{mn} \end{pmatrix} \\ = \begin{pmatrix} ca_{11} & ca_{12} & \dots & ca_{1n} \\ ca_{21} & ca_{22} & \dots & ca_{2n} \\ \vdots & \vdots & \ddots & \vdots \\ ca_{m1} & ca_{m2} & \dots & ca_{mn} \end{pmatrix}. \end{aligned}$$

In scalar multiplication, each entry of the matrix is multiplied by the scalar c. Consider, for example, multiplying the following 2×3 matrix over the integers by the integer 3:

$$3 \begin{pmatrix} 1 & 2 & 3 \\ 4 & 5 & 6 \end{pmatrix} = \begin{pmatrix} 3 & 6 & 9 \\ 12 & 15 & 18 \end{pmatrix}.$$

We can use scalar multiplication to find a vector whose length is a multiple of another vector. For example, if $V = (1\ 0\ 3\ 2)$, then $2V = (2\ 0\ 6\ 4)$ will be twice as long as V. We leave the idea of vector length and showing this fact in general to the Exercises.

We now take a look at some common applications of matrices.

Systems of Equations

Linear equations and linear inequalities play an important role in many fields of study. It is often possible to express a variety of relationships by writing those relationships in the form of one or more equations or inequalities to be considered as a system. For example, we might be told that the sum of the ages of two people is 49 and that the larger age minus the smaller is 13. To find out what the two ages are, we might represent them as x_1 and x_2, x_1 being the older. We can write two relationships for x_1 and x_2:

$$x_1 + x_2 = 49$$
$$x_1 - x_2 = 13.$$

There are several ways to solve this system of equations. By adding the equations, a single equation in x_1 is found: $2x_1 = 62$. From this equation x_1 is found to be 31. Substituting in either of the original equations shows that $x_2 = 18$. When the number of unknowns is small, direct manipulation of the equations is a reasonable and effective way to get solutions. Using elimination of a variable through addition or subtraction is one step commonly used, while another is that of substituting the solution of one variable in terms of another in an appropriate equation.

However, if the number of unknowns is large, even larger than two, these methods of elimination and substitution can be cumbersome. Moreover, we desire a more methodical technique for finding a solution.

Another way to look at this system of equations is to express the system in terms of a matrix equation $AX = C$, where the A stands for a matrix whose entries are the coefficients of the x_i, X stands for a column vector whose entries are the x_i, and C stands for a column vector with the constants as its entries. In the example about the ages we have

$$A = \begin{pmatrix} 1 & 1 \\ 1 & -1 \end{pmatrix}, X = \begin{pmatrix} x_1 \\ x_2 \end{pmatrix} \text{ and } C = \begin{pmatrix} 49 \\ 13 \end{pmatrix}.$$

So our matrix equation $AX = C$ is

$$\begin{pmatrix} 1 & 1 \\ 1 & -1 \end{pmatrix} \begin{pmatrix} x_1 \\ x_2 \end{pmatrix} = \begin{pmatrix} 49 \\ 13 \end{pmatrix}.$$

It is important to note that if we perform the matrix multiplication on the left of the equal sign, we get $\begin{pmatrix} x_1 + x_2 \\ x_1 - x_2 \end{pmatrix}$. The resulting

$$\begin{pmatrix} x_1 + x_2 \\ x_1 - x_2 \end{pmatrix} = \begin{pmatrix} 49 \\ 13 \end{pmatrix}$$

is equivalent to the two equations we started with.

We can write a matrix equation for any system of equations, but it may happen that not all unknowns appear in every equation. However, by filling in zero as the coefficients for unknowns not mentioned in a particular equation, we get an appropriate coefficient matrix whose dimensions are the number of equations by the number of unknowns.

For example, consider the equations

$$\begin{aligned} x_1 + x_3 - x_4 &= 12 \\ 2x_1 + x_4 &= 10 \\ 2x_2 + x_5 &= 0. \end{aligned}$$

We can express this as the matrix equation

$$\begin{pmatrix} 1 & 0 & 1 & -1 & 0 \\ 2 & 0 & 0 & 1 & 0 \\ 0 & 2 & 0 & 0 & 1 \end{pmatrix} \begin{pmatrix} x_1 \\ x_2 \\ x_3 \\ x_4 \\ x_5 \end{pmatrix} = \begin{pmatrix} 12 \\ 10 \\ 0 \end{pmatrix}.$$

✓ For the following system of equations, write the corresponding matrix equation:

$$\begin{aligned} 2x_2 - 4x_3 + 5x_4 &= 12 \\ 3x_1 - 3x_2 &= 10 \\ 5x_2 - 3x_3 + 2x_4 &= 0. \end{aligned}$$

Although there are several methods that can be used to solve systems of equations, we will focus on two particular methods, Cramer's rule and Gaussian elimination. Both methods provide a solution for a system of n equations in n unknowns, if a solution exists. Before stating Cramer's rule, there are two topics we need to investigate. One is a special function, called the *determinant*, that maps matrices to the set of their entries. The other is the question "Does every system of n equations in n unknowns have a solution?".

Let's consider existence of solutions. Suppose we have the following two equations in two unknowns:

$$\begin{aligned} 2x_1 + 3x_2 &= 12 \\ 4x_1 + 6x_2 &= 5. \end{aligned}$$

We note that the second equation has x_1 and x_2 coefficients that are twice the coefficients of x_1 and x_2, respectively, in the first equation. In order for both equations to be compatible, the constant in the second equation should be twice the constant in the first. If the 5 were replaced by 24, the two equations would be compatible and any value of x_1 and x_2 that satisfied one of the equations would also satisfy the other. But as the two equations stand, there are no values that can satisfy both of the equations at the same time. Therefore, this system of equations has no solution.

In this example it is easy to see the incompatibility. However, in general, if there are several equations in as many unknowns, it is not always obvious that the system is or is not consistent. By using the special function called the *determinant*, we will be able to find out two important facts about any system of n equations in n unknowns. First we will be able to tell whether or not a unique solution exists, and secondly we will be able to find the solution if it does exist. We note that some systems may have many solutions, rather than only one unique solution. We will not be looking at those systems here.

The Determinant

The special function, called the determinant, alluded to in the previous paragraphs is a function that takes a square matrix and assigns to it a scalar value in the set from which the matrix entries were drawn. For example, if the matrix has real numbers as entries, the determinant assigns a real number as the determinant value. If the matrix has entries from $\mathbb{Z}_7$, the determinant is a $\mathbb{Z}_7$ value.

Since there is an infinite number of sizes of square matrices, we will use a recursive approach to define the determinant. The base case is the one that defines the determinant value for a matrix that has only one row and one column (i.e., a single entry). In this case the determinant value is whatever that entry value is. Although the recursive part of the definition can now be made having the base case defined, before doing so, since the single entry matrices are not very revealing as far as finding the determinant, to be more instructive, we will show the determinant for matrices that have two rows and two columns. If

$$A = \begin{pmatrix} a_{11} & a_{12} \\ a_{21} & a_{22} \end{pmatrix}$$

we define the determinant of A, $\det A$, as

$$\det A = a_{11}a_{22} - a_{12}a_{21}$$

where the operations of multiplication (juxtaposition) and subtraction (or adding the additive inverse) are the scalar operations in the set of entries.

For example, if

$$A = \begin{pmatrix} 5 & 2 \\ 1 & 3 \end{pmatrix},$$

then $\det A = 5 \cdot 3 - 2 \cdot 1 = 13$.

Before proceeding to the general case of $n \times n$ matrices, let's look at the 3×3 case:

$$A = \begin{pmatrix} 5 & 3 & 0 \\ 4 & 2 & 1 \\ 3 & 0 & 1 \end{pmatrix}.$$

To get $\det A$, we choose any row of A. We then use each entry of the first row as a scalar, multiplying that entry times the determinant of the 2×2 matrix that results when the row and column containing the given entry in it is removed from A. If the sum of the subscripts for the given entry is even, we multiply the final answer by 1, and if the sum if odd, we multiply it by -1. To see how this works, we will apply this method to A.

Choosing the first row, 5 3 0, we take each entry and perform a scalar multiplication on the resulting 2×2 matrix as described:

$$5 \cdot \det \begin{pmatrix} 2 & 1 \\ 0 & 1 \end{pmatrix} + (-1) \cdot 3 \cdot \det \begin{pmatrix} 4 & 1 \\ 3 & 1 \end{pmatrix} + 0 \cdot \det \begin{pmatrix} 4 & 2 \\ 3 & 0 \end{pmatrix}.$$

Determinants for the resulting 2×2 matrices can be found the way we showed previously:

$$5 \cdot (2 \cdot 1 - 1 \cdot 0) - 3 \cdot (4 \cdot 1 - 1 \cdot 3) + 0 \cdot (4 \cdot 0 - 2 \cdot 3) = 5(2) - 3(1) + 0 = 7.$$

If we have a 4×4 matrix, we first split it into 3×3 matrices according to the preceding description. But to find the determinant of these 3×3 matrices we need to split each into

2×2 matrices and find their determinants. Notice that if we start with, say, 10×10 matrices, we have to keep splitting the matrices into smaller and smaller matrices. We can express this idea with the following recursive formula:

$$\det A = \sum_{j=1}^{n} (-1)^{1+j} a_{1j} \det A_{1j}$$

where A_{1j} means the matrix obtained by removing the first row and jth column from the matrix A, leaving a matrix with one fewer rows and one fewer columns. The base case here is $n = 2$, where the determinant is calculated as given previously. We have chosen to use the first row for finding the determinant, but we could have chosen any row. In the case of choosing row i, simply replace the 1 in the formula with i.

✓ Find the determinant of $A = \begin{pmatrix} 0 & 3 & 0 & 1 \\ 1 & 2 & 1 & -2 \\ 2 & 0 & 1 & -1 \\ 3 & 0 & -4 & -2 \end{pmatrix}$.

The determinant function has many applications in a variety of areas. We will use it for solving systems of equations where we have as many equations as we have unknowns. There is a theorem that tells when a solution exists and how to find it when it does. The theorem is called Cramer's rule.

Cramer's rule says that if $\det A \neq 0$, then the solution of the system of linear equations $AX = B$ is given by

$$x_i = \frac{\Delta_i}{\Delta}, \qquad \text{where } i = 1, 2, \ldots, n,$$

where $\Delta = \det A$ and $\Delta_i = \det A^i$, where A^i is the matrix obtained by replacing the ith column of A by B.

Note that if $\det A = 0$ we can not use Cramer's rule. Indeed, if $\det A = 0$ then the system of linear equations does not have a unique solution. Thus either the system has no solutions (that is, the equations are incompatible) or there is an infinite number of solutions.

Let's return to the example of a system of equations about ages that we solved earlier:

$$x_1 + x_2 = 49$$
$$x_1 - x_2 = 13.$$

We used a method of eliminating unknowns to solve the system, but such a method does not lend itself well to a computer program. Cramer's rule makes writing an algorithm much simpler. Applying Cramer's rule, we get

$$\Delta = \det \begin{pmatrix} 1 & 1 \\ 1 & -1 \end{pmatrix} = (-1 - 1) = -2$$

$$\Delta_1 = \det \begin{pmatrix} 49 & 1 \\ 13 & -1 \end{pmatrix} = (-49 - 13) = -62$$

$$\Delta_2 = \det \begin{pmatrix} 1 & 49 \\ 1 & 13 \end{pmatrix} = (13 - 49) = -36$$

So we get

$$x_1 = \frac{-62}{-2} = 31 \qquad \text{and} \qquad x_2 = \frac{-36}{-2} = 18.$$

✓ Use Cramer's rule to solve the following system of equations:

$$2x_1 + 5x_2 = 41$$
$$8x_1 - 3x_2 = 3.$$

Gaussian Elimination

Our second method for solving systems of equations is called Gaussian elimination. Intuitively, this method applies appropriate operations on rows of the coefficient matrix to result in transforming the matrix into the multiplicative identity for that matrix. Since we can express a system of equations as a matrix equation, $AX = C$, we note that if the A matrix happened to be the identity, then the values in C would be the solutions for the unknowns, respectively. As before, we will concern ourselves with square matrices only. For example, if we have

$$\begin{pmatrix} 1 & 0 \\ 0 & 1 \end{pmatrix} \cdot \begin{pmatrix} x_1 \\ x_2 \end{pmatrix} = \begin{pmatrix} 3 \\ 2 \end{pmatrix}$$

it is easy to see that upon performing the matrix multiplication, we get the result that $x_1 = 3$ and $x_2 = 2$. Our goal will be to transform a given coefficient matrix to the identity form without changing the relationships expressed in the original form.

Let's consider the following system of equations

$$4x_1 - x_2 = 10$$
$$x_1 + 3x_2 = 9.$$

The associated matrix equation is

$$\begin{pmatrix} 4 & -1 \\ 1 & 3 \end{pmatrix} \cdot \begin{pmatrix} x_1 \\ x_2 \end{pmatrix} = \begin{pmatrix} 10 \\ 9 \end{pmatrix}.$$

The matrix operations we will use in Gaussian elimination mimic operations we apply to equations. We'll first look at the equation operations and then move on to matrices. Examining the two equations, we note that if we were to interchange the two equations, we would get a coefficient of 1 for x_1 in the first equation. That exchange would obviously not in any way affect the values of x_1 and x_2. This gives us the equations:

$$\begin{aligned} x_1 + 3x_2 &= 9 \\ 4x_1 - x_2 &= 10. \end{aligned}$$

We could now multiply the first equation through by -4 without changing the unknown values and then add the result to the second equation, causing the coefficient of x_1 in the second equation to be 0 and eliminating the x_1 from that equation. This gives us the following:

$$\begin{aligned} x_1 + 3x_2 &= 9 \\ -13x_2 &= -26. \end{aligned}$$

We can now multiply the second equation by $-1/13$ (the multiplicative inverse of -13) to get

$$\begin{aligned} x_1 + 3x_2 &= 9 \\ x_2 &= 2. \end{aligned}$$

Note that now the coefficient of x_2 in the second equation is 1. Finally, we add -3 times the second equation to the first to get

$$\begin{aligned} x_1 &= 3 \\ x_2 &= 2. \end{aligned}$$

We have just seen three operations we can perform on systems of equations without changing the values of the unknowns: interchanging any two equations, adding a multiple of one equation to another, and multiplying any equation by a fixed value.

Analogously, we can perform the same kind of operations on the matrix form of the system of equations. We call these operations row operations. Since the column vector of unknowns is simply used as a place holder, we can ignore that part, looking only at what we call the augmented matrix. For the preceding system, the augmented matrix is

$$\left(\begin{array}{cc|c} 4 & -1 & 10 \\ 1 & 3 & 9 \end{array}\right).$$

Performing a sequence of row operations on the augmented matrix, we transform it to a new form in which the left part is the identity and the right part holds the solution to our equations. In this example, a good first step is to interchange the two rows:

$$\begin{pmatrix} 1 & 3 & | & 9 \\ 4 & -1 & | & 10 \end{pmatrix}.$$

Next we multiply the first row through by -4 and add it to the second row:

$$\begin{pmatrix} 1 & 3 & | & 9 \\ 0 & -13 & | & -26 \end{pmatrix}.$$

In the second row we would like to have a 1 where the -13 is and so we multiply the second row by the multiplicative inverse of -13:

$$\begin{pmatrix} 1 & 3 & | & 9 \\ 0 & 1 & | & 2 \end{pmatrix}.$$

We almost have the identity matrix on the left, but we need one more step to get it. We will multiply the second row by -3 and add it to the first row. Note that to do so will not spoil having a 1 in the first entry:

$$\begin{pmatrix} 1 & 0 & | & 3 \\ 0 & 1 & | & 2 \end{pmatrix}.$$

We call this the row reduced echelon form of the matrix. On the left we have the identity matrix. By converting back to the equation form, we get

$$\begin{pmatrix} 1 & 0 \\ 0 & 1 \end{pmatrix} \cdot \begin{pmatrix} x_1 \\ x_2 \end{pmatrix} = \begin{pmatrix} 3 \\ 2 \end{pmatrix}.$$

Matrix multiplication reveals that $x_1 = 3$ and $x_2 = 2$. We notice that upon completion of our row reduction, we get the identity as the left part of the augmented matrix. Now you should try this method on the two equations given earlier, for which there was no solution, and see what happens. Do you see why the augmented matrix you end up with(or rather the equations that correspond to this matrix) is inconsistent?

Intuitively, this method of solution is equivalent to having multiplied the original matrix of coefficients by its multiplicative inverse. Indeed, a system of equations without a unique solution will be one whose coefficient matrix has no inverse. In fact, there is a method for getting the multiplicative inverse of a given square matrix, if one exists, using row reduction.

Computing Multiplicative Inverses

For square matrices we know that there is a multiplicative inverse. For $n \times n$ matrices it is

$$I = \begin{pmatrix} 1 & 0 & \dots & 0 \\ 0 & 1 & \dots & 0 \\ \vdots & \vdots & \ddots & \vdots \\ 0 & 0 & \dots & 1 \end{pmatrix}.$$

It turns out that I commutes under multiplication with all other matrices of the same dimension, even though, in general, matrix multiplication is not commutative.

Recall that the *multiplicative inverse* (or just *inverse*) of the (square) matrix A is the matrix denoted A^{-1} with the property that $AA^{-1} = I$. Two facts follow from this: It's also true that $A^{-1}A = I$ even though matrix multiplication is not in general commutative, and if A^{-1} exists it is unique. Not all square matrices have inverse. For instance, the zero matrix (one with all zero entries) obviously does not have an inverse since if we multiply any matrix by the zero matrix, we'll get the zero matrix. But there are other examples that do not have inverses, such as one with one or more zero rows.

Let's turn our attention to computing the inverse of a matrix. Consider the coefficient matrix A in the example of a system we solved using Cramer's rule:

$$A = \begin{pmatrix} 1 & 1 \\ 1 & -1 \end{pmatrix}.$$

Note the importance of finding A^{-1}. If we knew the multiplicative inverse of A, then we could solve the system by multiplying both sides of the matrix equation $AX = C$ by A^{-1}, yielding $A^{-1}C = A^{-1}AX = IX = X$.

If

$$A^{-1} = \begin{pmatrix} x_1 & x_2 \\ x_3 & x_4 \end{pmatrix},$$

then $AA^{-1} = I$. That is,

$$\begin{pmatrix} 1 & 1 \\ 1 & -1 \end{pmatrix} \cdot \begin{pmatrix} x_1 & x_2 \\ x_3 & x_4 \end{pmatrix} = \begin{pmatrix} 1 & 0 \\ 0 & 1 \end{pmatrix}.$$

This gives us four equations in four unknowns. We could use Gaussian elimination to find the solution. Instead of displaying these four equations on four lines, we'll modify the augmented matrix so the right side starts as the identity matrix and perform row operations on A and the identity matrix at the same time:

$$\begin{pmatrix} 1 & 1 & | & 1 & 0 \\ 1 & -1 & | & 0 & 1 \end{pmatrix}.$$

We'll skip the intermediate steps here. (You should do this.) After performing the necessary row operations to transform the left side to the identity matrix, we get

$$\begin{pmatrix} 1 & 0 & | & \frac{1}{2} & \frac{1}{2} \\ 0 & 1 & | & \frac{1}{2} & -\frac{1}{2} \end{pmatrix}.$$

Consider the final matrix on the right side. This is the multiplicative inverse of A. To confirm this, we multiply as follows:

$$\begin{pmatrix} \frac{1}{2} & \frac{1}{2} \\ \frac{1}{2} & -\frac{1}{2} \end{pmatrix} \begin{pmatrix} 1 & 1 \\ 1 & -1 \end{pmatrix} = \begin{pmatrix} \frac{1}{2} + \frac{1}{2} & \frac{1}{2} - \frac{1}{2} \\ \frac{1}{2} - \frac{1}{2} & \frac{1}{2} + \frac{1}{2} \end{pmatrix} = \begin{pmatrix} 1 & 0 \\ 0 & 1 \end{pmatrix}.$$

As stated before, it turns out that if a square matrix does have a multiplicative inverse, that inverse is unique and serves as both a left inverse and a right inverse. So, if we are given matrix A and we find A^{-1}, then $AA^{-1} = A^{-1}A = I$. Check this for the preceding example.

✓ Find the multiplicative inverse of $A = \begin{pmatrix} 2 & 1 \\ 3 & -2 \end{pmatrix}$.

We've seen that the zero matrix doesn't have an inverse, but others do not also. For example, consider the 2×2 matrix

$$A = \begin{pmatrix} 2 & 0 \\ 3 & 0 \end{pmatrix}.$$

To see if A has an inverse, we augment A with I and perform the appropriate row operations:

$$\left(\begin{array}{cc|cc} 2 & 0 & 1 & 0 \\ 3 & 0 & 0 & 1 \end{array}\right).$$

After performing the row operations, we see we do not end up with the identity on the left side. Indeed, we end with

$$\left(\begin{array}{cc|cc} 1 & 0 & \frac{1}{2} & 0 \\ 0 & 0 & -\frac{1}{2} & \frac{1}{3} \end{array}\right).$$

If we transform this back to matrix equation form, we get

$$\begin{pmatrix} 1 & 0 \\ 0 & 0 \end{pmatrix} \begin{pmatrix} x_1 & x_2 \\ x_3 & x_4 \end{pmatrix} = \begin{pmatrix} \frac{1}{2} & 0 \\ -\frac{1}{2} & \frac{1}{3} \end{pmatrix}.$$

This multiplication yields four equations:

$$x_1 = \frac{1}{2}, \quad x_3 = 0, \quad 0 = -\frac{1}{2}\,, \text{ and } \quad 0 = \frac{1}{3}.$$

The last two statements are obviously false under any circumstances. This means that our system of equations has no solution and so our original matrix A has no multiplicative inverse. There are many other 2×2 matrices without inverses.

Row reducing a matrix to its echelon form not only gives us a methodical approach to finding a multiplicative inverse for the matrix, but it also allows us to tell whether a multiplicative inverse exists. If the row reduced form turns out not to be the identity, but rather a matrix that looks like the identity for the first few rows but has all 0's for the final one or more rows, then we can conclude that the matrix in question does not have a unique multiplicative identity.

Encryption Revisited

Now that we know two methods for solving systems of equations, we are ready to look at one more way to encrypt messages, a method that employs a system of congruences with as many congruences as there are unknowns. Since all congruences, such as $3x \equiv 5 \bmod n$, are known to have a solution when the modulus, n, is prime, we will choose a prime for the congruences in our system.

In choosing the prime we will take into consideration the number of symbols we may want to encrypt. For example, suppose we want to send messages that consist of the 26 alphabetic letters, say all capitals, rather than needing 52 symbols to accommodate both capitals and lowercase. We could associate the letters, $A, B, \ldots, Z$ with $1, 2, \ldots, 26$, respectively. The smallest prime larger than 26 is 29, so we select the $\mathbb{Z}_{29}$ system with its usual addition and multiplication. Since 29 is prime, all the nonzero elements have multiplicative inverses and so any congruence has a unique solution.

Suppose we want to encrypt the word `HELLO`. We will choose a system of congruences mod 29 as our key, the number of congruences determining how many letters we can encrypt at once. The method used here is called a *digraphic* cipher because we'll encrypt a pair of letters at a time. (A pair of letters is called a *digram.*) So we will need two congruences. We need to pick the congruences so that they are not just multiples of each other, for the reasons we saw in the previous section.

We could choose the coefficients of our unknowns at random. Let p_1 and p_2 be the pair of plaintext letters and c_1 and c_2 be the resulting ciphertext letters. Suppose we choose the following set of congruences:

$$2p_1 + 3p_2 \equiv c_1$$
$$5p_1 + 25p_2 \equiv c_2.$$

We're doing all our arithmetic here in $\mathbb{Z}_{29}$, so we'll use equal signs in place of equivalence signs to simplify the notation slightly. Note that all the coefficients can be positive. However, it can simplify the arithmetic if we write the coefficient 25 as -4, since 25 is the additive inverse of 4. We can now write the system in a matrix form:

$$\begin{pmatrix} 2 & 3 \\ 5 & -4 \end{pmatrix} \begin{pmatrix} p_1 \\ p_2 \end{pmatrix} = \begin{pmatrix} c_1 \\ c_2 \end{pmatrix}.$$

To encrypt, we will simply fill in the numbers associated with the letters in our message, putting the number for the first letter in place of p_1 and the number for the second letter in place of p_2. We first encrypt `HE` from `HELLO`. `H` corresponds to 8, while `E` corresponds to 5. Filling in the first equation using 8 for p_1 and 5 for p_2, we get $2 \cdot 8 + 3 \cdot 5 = 31 \bmod 29 = 2$. 2 corresponds to `B`. Similarly, the second equation yields $5 \cdot -4 \cdot 5 = 20 \bmod 29$. 20 is associated with `S`. Hence the digram `HE` is encrypted as `BS`.

Of course, it is possible to get some number other than those between 1 and 26, namely 0, 27, or 28. To complete the association of symbols and numbers, let's use a space for 0, a comma for 27, and a period for 28. This allows us to put in a symbol for any resulting integer mod 29. Having encrypted the first two letters, we are ready to continue the encryption process.

The next pair of letters in our plaintext is `LL`. It is interesting to note that even though the next two letters are both the same, `L`, they will encrypt to different letters because the two congruences yield different constants when the values are filled in. This fact, that the same letter may encrypt to different letters in different parts of the message, makes such a scheme particularly difficult to decrypt.

After encrypting `LL`, as are left with one letter, `O`, to encrypt. Since our encryption method requires two characters to encrypt, we simply choose another character to pair with `O`. This could be a space, for example. For security reasons, we don't want to be predictable in our choice of this extra letter, so we pick one at random. Here, let's pair `O` with `X`.

✓ Encrypt the diagrams `LL` and `OX` using the previous encryption matrix .

The decryption process also processes two letters at a time. The purpose of decryption is to recover the plaintext, p_1 and p_2, from the ciphertext, c_1 and c_2. But if E is the encryption matrix, the encryption equation is

$$E\begin{pmatrix} p_1 \\ p_2 \end{pmatrix} = \begin{pmatrix} c_1 \\ c_2 \end{pmatrix}.$$

But then, if we multiply both sides of the equation by the matrix E^{-1}, we get

$$E^{-1}E\begin{pmatrix} p_1 \\ p_2 \end{pmatrix} = E^{-1}\begin{pmatrix} c_1 \\ c_2 \end{pmatrix},$$

which implies that

$$I\begin{pmatrix} p_1 \\ p_2 \end{pmatrix} = E^{-1}\begin{pmatrix} c_1 \\ c_2 \end{pmatrix},$$

and so

$$\begin{pmatrix} p_1 \\ p_2 \end{pmatrix} = E^{-1} \begin{pmatrix} c_1 \\ c_2 \end{pmatrix}.$$

Thus the decryption matrix is E^{-1}. So our task here is to compute the inverse of our encryption matrix. (Keep in mind that we are doing arithmetic in $\mathbb{Z}_{29}$.) Proceeding as before

$$\left(\begin{array}{cc|cc} 2 & 3 & 1 & 0 \\ 5 & -4 & 0 & 1 \end{array}\right).$$

First we multiply the first row by 1/2. But here (in $\mathbb{Z}_{29}$) 1/2 is the multiplicative inverse of 2, which by trial and error is 15, since $2 \cdot 15 = 30 = 1 \bmod 29$. Doing this, our augmented matrix becomes

$$\left(\begin{array}{cc|cc} 1 & 16 & 15 & 0 \\ 5 & -4 & 0 & 1 \end{array}\right).$$

Adding -5 times the first row to the second yields

$$\left(\begin{array}{cc|cc} 1 & 16 & 15 & 0 \\ 0 & 3 & 12 & 1 \end{array}\right).$$

Multiplying the third row by 1/3 (which is 10, as you can easily see) gives us

$$\left(\begin{array}{cc|cc} 1 & 16 & 15 & 0 \\ 0 & 1 & 4 & 10 \end{array}\right).$$

Finally, adding -16 times the second row to the first yields

$$\left(\begin{array}{cc|cc} 1 & 0 & 9 & 14 \\ 0 & 1 & 4 & 10 \end{array}\right).$$

Thus E^{-1}, our decryption matrix is

$$\begin{pmatrix} 9 & 14 \\ 4 & 10 \end{pmatrix}.$$

We encrypted HE and got BT. If we decrypt BT we should recover HE. Since B corresponds to 2 and T to 20

$$\begin{pmatrix} 9 & 14 \\ 4 & 10 \end{pmatrix} \begin{pmatrix} 2 \\ 20 \end{pmatrix} = \begin{pmatrix} 18 + 280 \\ 8 + 200 \end{pmatrix} = \begin{pmatrix} 298 \\ 208 \end{pmatrix} = \begin{pmatrix} 8 \\ 5 \end{pmatrix}.$$

Since 8 corresponds to H and 5 to E, it works!

> ✓ Decrypt the diagrams you calculated above when encrypting LL and OX. You should recover your plaintext digrams, of course.

Entire volumes are written on the theory and applications of matrices. Here, we have concentrated on the basic operations on matrices, illustrating their uses with a few example applications. The Exercises will give you a chance to test yourself on your understanding of matrices.

Exercises

Use these matrices for the first seven exercises.

$$A = \begin{pmatrix} 5 & 6 & 4 & 2 \\ 1 & 7 & 3 & 2 \\ 3 & 1 & 0 & 3 \\ 5 & 0 & 0 & 1 \end{pmatrix} \qquad B = \begin{pmatrix} 1 & 2 & 3 & 0 \\ 4 & 3 & 0 & 1 \\ 0 & 0 & 1 & 1 \end{pmatrix}$$

$$C = \begin{pmatrix} 3 & 1 & 1 & 1 \\ 2 & 0 & 0 & 8 \\ 0 & 8 & 9 & 0 \\ 0 & 1 & 3 & 3 \end{pmatrix} \qquad D = \begin{pmatrix} 1 & 1 & 0 & 1 \end{pmatrix}$$

$$E = \begin{pmatrix} 7 & 6 & 4 & 2 \\ 0 & 0 & 3 & 1 \\ 3 & 2 & 4 & 0 \end{pmatrix} \qquad F = \begin{pmatrix} 2 & 0 \\ 3 & 1 \end{pmatrix}$$

1. What are the dimensions of the matrices?

2. Which pairs of the given matrices can be added?
3. Which pairs can be multiplied?
4. Find all the sums defined.
5. Find all the products defined.
6. Find inverses for any of the matrices that have them.
7. Find determinants for any of the matrices for which a determinant is defined.
8. Let

$$A = \begin{pmatrix} 2 & 4 & 1 \\ 1 & 2 & 5 \\ 3 & 1 & 6 \end{pmatrix} \text{ and } B = \begin{pmatrix} 1 & 3 & 5 \\ 2 & 3 & 4 \\ 0 & 6 & 0 \end{pmatrix}.$$

   a. Assuming the entries are integers, find $A + B$.
   b. Assuming the entries are integers, find $3 \cdot A$.
   c. Assuming the entries are integers, find $A \cdot B$.
   d. Assuming the entries are integers, find $\det A$
   e. Assuming the entries are rationals, find the multiplicative inverse of A.
   f. Assuming the entries are from $\mathbb{Z}_7$, follow the directions of a, b, c, d, and e.

9. Using Cramer's rule, solve the following system of equations, assuming the coefficients are integers.

   a.

$$x_1 + x_2 + x_3 = 6$$
$$x_1 - x_2 + x_3 = 2$$
$$x_1 + 2x_2 + 3x_3 = 14$$

   b. Solve the same equations by Gaussian elimination.
   c. Why can't the following be solved using Cramer's rule?

$$2x_1 - 3x_2 + x_3 = 4$$
$$3x_1 + 2x_2 - x_3 = 9$$
$$x_1 + 5x_2 - 2x_3 = 2$$

d. Solve this system using Gaussian elimination (if there are any solutions).
e. Why can't the following be solved using Cramer's rule?

$$4x_1 + x_2 - 5x_3 = 13$$
$$2x_1 - 3x_2 + x_3 = 7$$
$$x_1 + 2x_2 - 3x_3 = 3$$

f. Solve this system using Gaussian elimination (if there are any solutions).

10. Solve the system in Exercise 9a, assuming the entries are from $\mathbb{Z}_{11}$.
11. Find the inverse in $\mathbb{Z}_{29}$ of the encryption matrix

$$\begin{pmatrix} 1 & 3 \\ 5 & 1 \end{pmatrix}.$$

12. Encrypt `SUMMER` using the encryption matrix given in Exercise 11.
13. Now decrypt the message you got in Exercise 12, using the inverse of the encryption matrix. (You should recover `SUMMER`, of course.)
14. Decrypt `KVJY` if this is the ciphertext encrypted by the encryption matrix used in the text.
15. Suppose you want to use a trigraphic encryption scheme in $\mathbb{Z}_{29}$. Use the following system of congruences:

$$\begin{aligned} 2p_1 + 3p_2 + p_3 &\equiv c_1 \\ 5p_1 + 25p_2 + 4p_3 &\equiv c_2 \, . \\ p_1 + 22p_2 + 7p_3 &\equiv c_3 \end{aligned}$$

(Note that this encrypts three letters (a trigram) at a time.) Encrypt the message `HELLO`. (Pad this message with a blank.)

16. Using this same system as in Exercise 15, find the decryption matrix. (This will test your calculating skills in $\mathbb{Z}_{29}$.) Check your result on the ciphertext from the previous exercise.

17. Find det A, where

$$A = \begin{pmatrix} 0 & 2 & 1 & 2 \\ 1 & 0 & 0 & 0 \\ 2 & 3 & 0 & 0 \\ -1 & 2 & 0 & 1 \end{pmatrix}.$$

Recall that you can expand on any row. Make your work as easy as possible by expanding on the row that has the most 0 entries.

18. Find det A, where

$$A = \begin{pmatrix} 1 & 1 & 1 \\ 1 & 1 & 1 \\ 1 & 1 & 1 \end{pmatrix}.$$

19. Find det A, where

$$A = \begin{pmatrix} 1 & 0 & 0 \\ 0 & 1 & 0 \\ 0 & 0 & 1 \end{pmatrix}.$$

20. Find det A, where

$$A = \begin{pmatrix} 0 & 0 & 1 \\ 0 & 1 & 0 \\ 1 & 0 & 0 \end{pmatrix}.$$

21. Suppose $AB = I$, where A, B, and I are all square matrices of the same size and I is the identity matrix. Thus $B = A^{-1}$. Show that $BA = I$ also.

22. Show that A^{-1} is unique. That is, show that if $AB = I$ and $AC = I$ then $B = C$. (Use the fact from the Exercise 21.)
23. Verify, for 2×2 matrices, that I commutes with all matrices.
24. Here's a nice little formula for computing the inverse of a 2×2 matrix:

$$\begin{pmatrix} a & b \\ c & d \end{pmatrix}^{-1} = \begin{pmatrix} \frac{d}{ad-bc} & \frac{-b}{ad-bc} \\ \frac{-c}{ad-bc} & \frac{a}{ad-bc} \end{pmatrix}.$$

Verify this. Note that if $ad-bc = 0$, the inverse does exist. But $ad - bc$ is the determinant of the matrix. This is a general fact: The inverse of a square matrix exists if and only if its determinant is nonzero.

25. Show that matrix multiplication distributes over matrix addition.

Programming Problems

1. Write a program that adds, multiplies, and performs scalar multiplication on 3×3 matrices with integer entries.
2. Write a program that adds, multiplies, and performs scalar multiplication on square matrices with up to 20 rows and columns.
3. Design and implement a matrix abstract data type (ADT) (in C++ use a template class) that permits the declaration of square matrices (in C++ this is a constructor), addition, multiplication, and scalar multiplication of square matrices of any dimension.
4. Write a program that computes the determinant for 2×2 matrices.
5. Write a program that computes the determinant for 3×3 matrices.

* 6. Write a program that computes the determinant for any square matrix. *Hint*: Recursion may be useful here. If you succeed with this program, insert the function you have written into the ADT you wrote in Problem 3.

7. Write a program to use Cramer's rule to solve systems of equations with integer coefficients in two unknowns. *Hint*: You might use your determinant function from Problem 4.

8. Write a program to use Cramer's rule to solve systems of equations with integer coefficients in three unknowns.

9. Write a program to use Cramer's rule to solve systems of equations with coefficients from $\mathbb{Z}_{29}$ in two unknowns.

10. Write a program to use Cramer's rule to solve systems of equations with coefficients from $\mathbb{Z}_{29}$ in three unknowns.

11. Write a program to find the multiplicative inverse of a 2×2 matrix. Try the same for a 3×3 matrix.

* 12. Write a program to find the multiplicative inverse of any square matrix with integer coefficients (if the matrix has one). You should write defensive code that checks first to see if the matrix has an inverse.

13. Write a program that inputs an encryption matrix (as in the text) and a message and outputs the decrypted message. Have your program check that the encryption matrix does indeed have an inverse.

* 14. Write a program that inputs a matrix, then performs Gaussian elimination on this matrix and outputs the result. (This program is challenging. You may have to exchange rows if a coefficient is zero in the wrong place.)

Chapter 10
Graphs

Some of the most famous problems in computer science are based in an area of mathematics known as graph theory. Because there are so many such problems and because they are about such a wide variety of topic areas, it is both difficult and misleading to choose only a few of those problems to represent the broad collection. For this reason we will not even attempt to be representative or comprehensive. Rather, our goal here is to introduce some basic concepts of graph theory, along with a few examples, so that the reader will have the background necessary to read and understand problems about graphs.

We begin with a problem that is well-known, practical in nature, and solvable by some efficient algorithms. In fact, our problem has been addressed so often that it has a name: "the Euler circuit problem." The problem is named for the mathematician of the eighteenth century whose work is well known is many areas of mathematics. In fact, we trace graph theory to Euler.

Euler Circuits and Tours

It would not be surprising if at some time while you were in elementary school, you were challenged to draw the following figures without raising your pencil from the paper.

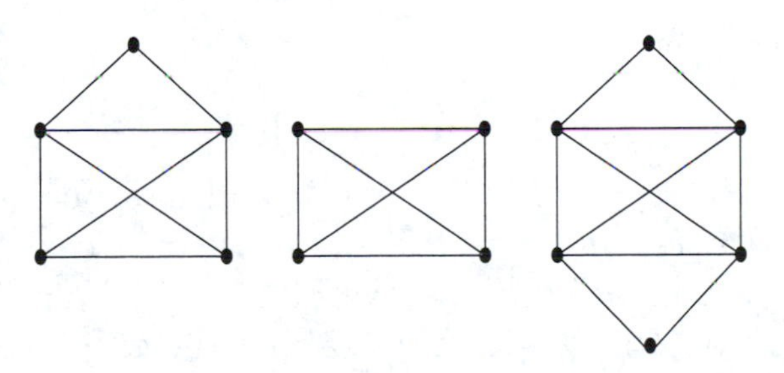

Another way to state this problem is as follows: Find a path that visits every edge in the graph exactly once. This statement is called "the Euler tour problem." A path that visits every edge in the graphs exactly once and returns to the starting vertex is known as "an Euler circuit."

Originally, Euler introduced this problem by asking whether or not it would be possible to visit every bridge in Konigsburg (now Kaliningrad) exactly once, returning to the starting point. The city had seven bridges connecting the two sides of a river and the two islands. The bridge problem is one instance of the general Euler tour problem. The following is a drawing of the bridges.

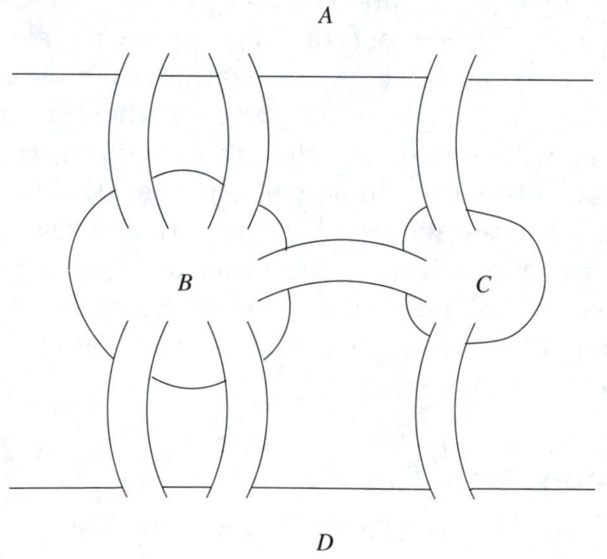

Before finding a solution to this problem, if one exists, we will have a look at some concepts and terms used when describing graphs.

Symbols and Terms for Graphs

A *graph* is a pair of finite sets, V and E, where V is called the set of *vertices* and E is called the set of *edges*. Each edge consists of a pair of vertices. We often write $G = (V, E)$, an ordered pair of sets. We then talk about the graph G, meaning the sets V and E. If A and B are vertices and (A, B) is an edge, we often abbreviate the edge (A, B) as AB.

For example, suppose $V = \{A, B, C, D\}$ and $E = \{(A, B), (A, C), (C, D), (A, D), (B, C)\}$. Here $G = (V, E)$ is a graph with four vertices and five edges.

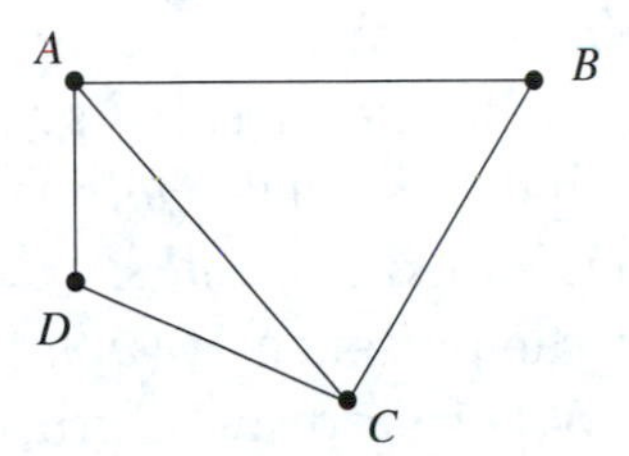

In a case where every vertex is connected to every other vertex in the graph, we have what is called a *complete graph*. For example, the complete graph on four vertices must have $\binom{4}{2=6}$ edges and could be drawn as follows:

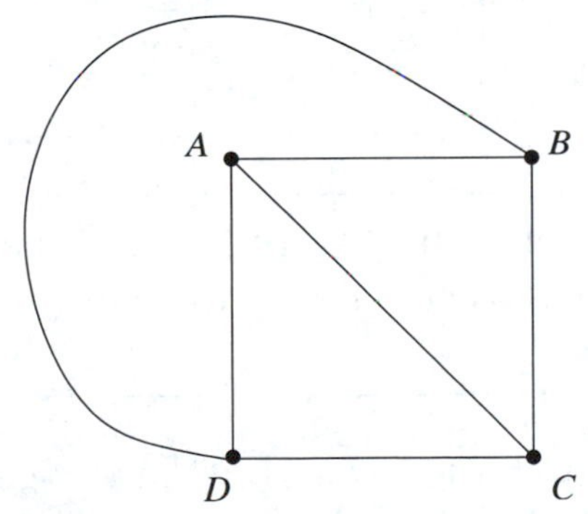

We have drawn the edge between B and D so that it does not cross the edge AC just to avoid the impression that another vertex might exist where BD would cross AC. In general, the complete graph on n vertices has $\binom{n}{2}$ edges.

✓ Draw all the graphs with four vertices.

The number of edges incident with a given vertex is called the *degree* of the vertex. For example, in the picture of the complete graph on four vertices, the degree of vertex A is 3, since edges AB, AC, and AD are all incident with vertex A.

A *path* in a graph is a sequence of vertices $v_1, v_2, \ldots, v_n$ such that (v_i, v_{i+1}) is in E for every i between 1 and $n - 1$. The *length* of the path from some v to w is the number of edges on the path. If all the vertices on a path are distinct, we say that the path is a *simple* path.

In some graphs, called *undirected* graphs, the presence of edge (v, w) implies that the edge (w, v) is also included. In *directed* graphs, often called *digraphs*, (v, w) is considered an ordered pair, so that the presence of (v, w) in E does not imply that (w, v) is in E. We often draw digraphs with arrows. For example,

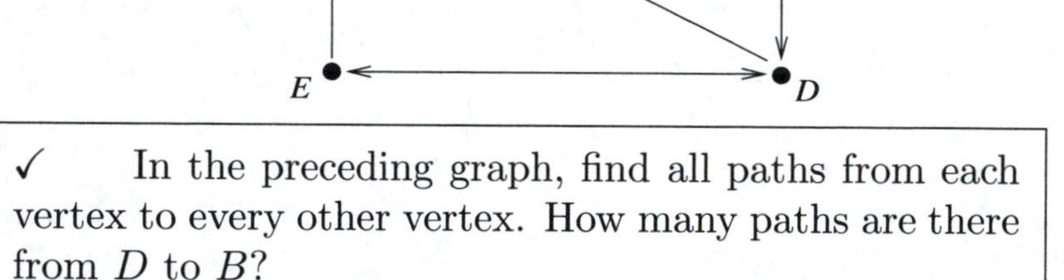

✓ In the preceding graph, find all paths from each vertex to every other vertex. How many paths are there from D to B?

In the picture both ED and DE are edges. AB is an edge, but not BA. Notice that there is a path from A back to A, namely AC, CD, DA. Such a path is called a *cycle*. In this case, since the cycle includes three edges, it is called a cycle of length 3. Another cycle in the graph is AC, CD, DE, EA, a cycle of length 4. A graph that has no cycles is called an *acyclic graph*.

In some graphs each edge may have a number associated with it. Such a graph is called a *weighted graph*. For example, a graph used to represent streets as edges and intersections as vertices may associate with each street a number of vehicles that can be accommodated at a given time on that street. Many traffic flow problems can be represented as graph problems.

One other concept important for understanding graph problems is the idea of connectivity. An undirected graph is said to be *connected* if there is a path from every vertex to every other vertex. Remember that a path is a sequence of edges, not just a single edge, so being connected is different from being complete.

When a directed graph has this same property, a path from every vertex to every other vertex, the graph is called *strongly connected.* If a directed graph is not strongly connected, but between every pair of vertices u and v there is either a path from u to v or from v to u (but perhaps not both), then the graph is called *weakly connected.*

A connected, acyclic graph is what we call a *tree.* A collection of trees is called a *forest.* For example, the following are trees:

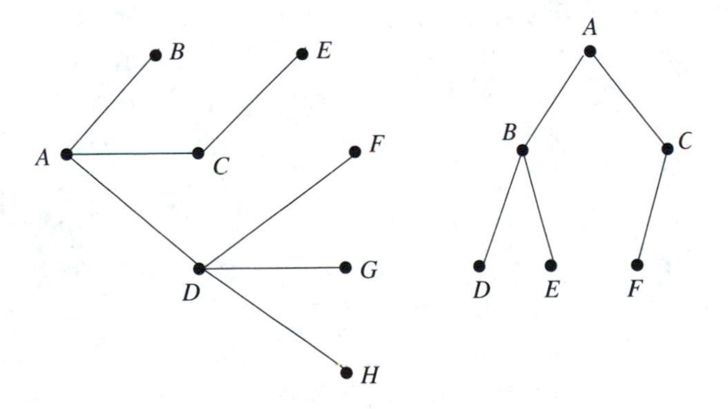

A Return to Euler Circuits

Armed with our new concepts and terms, we are ready to take another look at the Euler circuit and tour problems. Let's look first at the circuit problem, the one that requires us to visit every edge exactly once and return to the starting point. Since to get started it is necessary to choose an edge leaving the starting point, S, we can conclude that it would be impossible to return to S unless another edge also is incident with S. (Remember that we can visit each edge only once.) But every visit to a node other than S must have an entry edge and an exit edge. Thus, if a graph has an Euler cirucit, the degree of every vertex must be even. We can look at the three graphs from before (reproduced below) and conclude that neither the first nor the second can have an Euler circuit, since not all vertices have even degree.

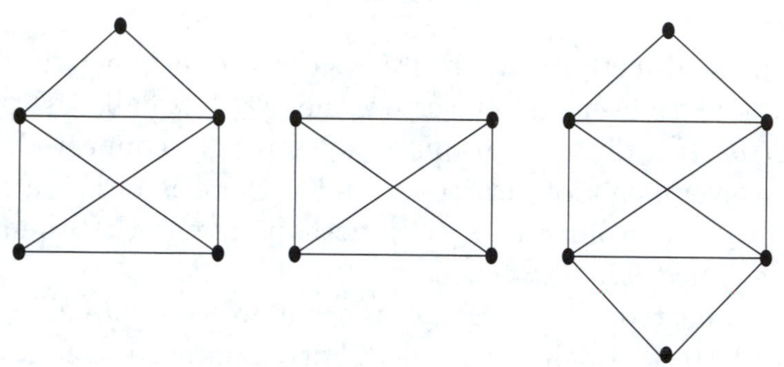

Amazingly, the converse is true. That is, if every vertex is even in a connected graph, then it has an Euler circuit. We will not prove that here, but the reader should find an Euler circuit in the third graph above and draw some other graphs whose vertices have even degree and see that they also have Euler circuits.

The requirements for a graph to have an Euler tour are slightly different from those for a circuit. But given that we know a connected graph has an Euler circuit if and only if each vertex has even degree, it is easy to prove that a connected graph has an Euler tour if and only if all vertices have even degree (in which case the tour is actually a circuit) or exactly two vertices have odd degree.

First, suppose a graph has an Euler tour that is not a circuit. Except for the starting and ending vertices, vertices have even degree, by the argument given about circuits. But the starting vertex has the edge initially used in the tour that is not matched with an edge entering the starting vertex and so it has odd degree. Similarly, we see that the ending vertex has odd degree. Conversely, suppose we have a connected graph with exactly two vertices of odd degree. Call these vertices S and E. Now add a new node, N, and the edges SN and EN. We now have a connected graph where all vertices have even degree and so this graph has an Euler circuit. If you remove the edges SN and NE from this circuit, you now have an Euler tour with end nodes S and E. This proves our claim.

When you study algorithms, you will discover that there is a simple, efficient solution to finding an Euler circuit or tour.

The Euler problems were about visiting edges of a graph exactly once. It is natural to pose the analogous question about visiting vertices: Can you visit each vertex of a particular graph exactly once? This problem also has a special name, the Hamiltonian circuit problem. The surprise is that although we were able to examine the indices of the vertices of a given graph and know without doubt whether hunting for an Euler tour or circuit was feasible, we cannot do the same for the Hamiltonian problem. The only known solution for the Hamiltonian problem is the one that requires us to try every possible path through the graph to see if any of them visit each vertex exactly once.

Minimal Spanning Tree

We have chosen one other graph problem to present here. The choice is based on several interesting facts. The problem, called the "minimal spanning tree problem," has some efficient solutions that are easy to understand without additional background. The problem also has some important applications that are obviously useful. Finally, the problem makes use of many of the fundamental concepts of graph theory.

This problem can be phrased in many ways, depending on the domain in which we wish to apply it. We have chosen the domain of weighted graphs, but you can easily find other domains in which the problem makes sense. Here is the problem:

We want to connect several sites with a network so that those sites can communicate electronically. We have performed an analysis on the sites and know how much it will cost to connect any given pair of sites. The costs vary according to circumstances. For example, some sites may be located in areas where making the connection is very cheap, such as sites within the same room, while others require great expense, such as those in remote locations. We can represent the situation by placing dots to represent the sites and connecting lines to represent the required connections. We can place the costs on the lines to indicate how much it would be to connect any particular pair of sites. What we have is a weighted graph where the vertices represent the sites and edges represent connection between those sites. The weights for the edge are the costs of the connections.

When we have found the cheapest way to allow all our vertices to communicate, the resulting graph will have our original set of vertices but only a subset of the edges. Moreover, the solution will be a tree, hence the name of the problem. No cycles will be allowed because any cycle would indicate some redundancy and hence would not be the cheapest way to connect the sites. We include one example to illustrate the situation:

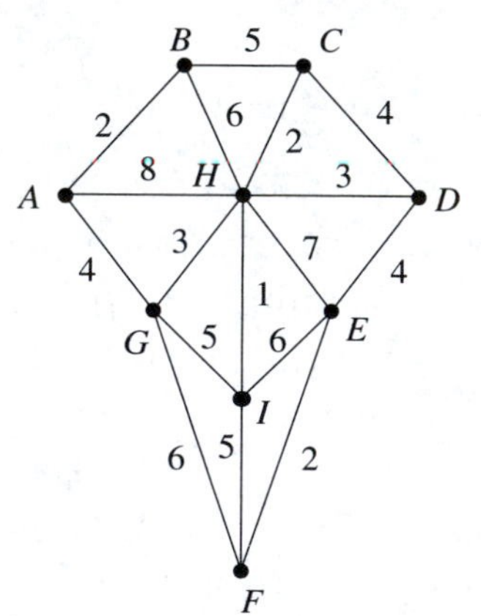

One way to find the minimal spanning tree is to use an algorithm made famous by Kruskal:

1. Sort all the edges by cost from smallest to largest.
2. *Repeat*

   Remove the next smallest edge from the collection of edges.

   If the vertices on that edge are not already connected, then add that edge to the spanning tree.

   Otherwise, discard the edge.

   Until there are no more edges.

At each step we choose the cheapest possible edge to connect two vertices that are not already connected. The result is that we never put in a redundant connection and we always pick the cheapest possible edge. Let's apply this algorithm to the graph we used to introduce the problem:

First we order the edges by cost:

$$HI:1,\ EF:2,\ AB:2,\ CH:2,\ DH:3,\ GH:3,\ CD:4,$$
$$AG:4,\ DE:4,\ GI:5,\ FI:5,\ BC:5,\ BH:6,\ EI:6,$$
$$FG:6,\ EH:7,\ AH:8.$$

Those edges with identical weights could be listed in an order, of course. The order could affect the minimal spanning tree that results from applying the algorithm. In general, a given weighted graph may have more than one minimal spanning tree.

Next, we repeatedly add edges to our tree. The critical condition is that we not add an edge between vertices already connected in the tree we've built so far. That is, we do not want to introduce a cycle. The first six edges added are HI, EF, AB, CH, DH, and GH. When we come to edge CD, we see that the tree we have so far constructed has C connected to D (C to H to D) and so we do not add edge CD to our tree. You are asked to finish this example in the Exercises.

Some Programming Considerations

Representing graphs in a computer program may be done is a variety of ways. We will look briefly here at two possibilities. The first is to use a two-dimensional array, say G. We think of the indices in the array as ordered pairs of vertices. In some languages the index set for arrays is not restricted to integers but can be any set. In such languages (Pascal, for example), it is possible to talk about $G[B, C]$ to access an entry in the array. For an unweighted graph, G, we store a 0 to mean there is no edge between the two given vertices, whereas a 1 means that there is an edge. For example, if there is an edge between B and C, then $G[B, C] = 1$.

If the graph is weighted, then each entry represents the weight of the edge. For example, if G is a weighted graph with an edge of cost 12 connecting vertex B to vertex C, then $G[B, C] = 12$. For vertices that have no edge connecting them, some particular fixed value is entered (such as -1), to indicate "no edge."

If we are using a language that permits only natural number indices (for example, C++), then we must represent all the vertices as integers, choosing some way of mapping the actual names of the vertices to the natural numbers.

A second common way of representing graphs is the use of what are called adjacency lists. In this approach there is a list for each vertex with entries for every vertex the given one is connected to. For example, if in some graph, G, there is an edge from A to C, another from A to E, and another from A to B, then the list for A would contain C, E, and B. There is no assumption made about the order, nor should there be any assumption made about the entries relative to each other. In other words, the fact that both C and E appear on A's list does not imply anything about an edge between C and E. To find out we would need to look at the list for C or for E.

To illustrate these two approaches, consider the following unweighted, undirected graph, G:

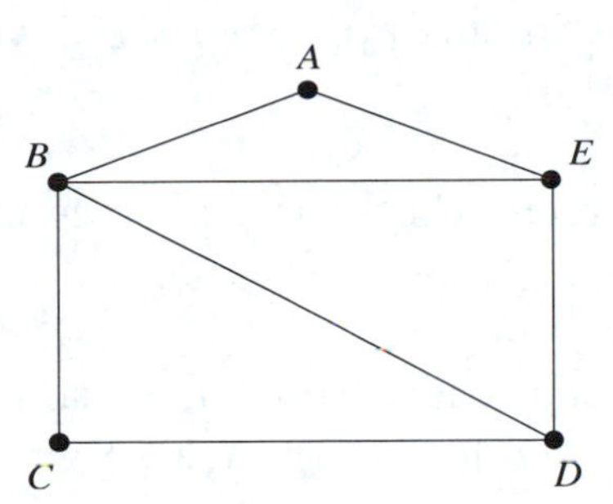

The two-dimensional array for G can be represented pictorially by

	A	B	C	D	E
A	0	1	0	0	1
B	1	0	1	1	1
C	0	1	0	1	0
D	0	1	1	0	1
E	1	1	0	1	0

The list approach is shown as follows:

$$A \longrightarrow B \longrightarrow E$$
$$B \longrightarrow A \longrightarrow E \longrightarrow C \longrightarrow D$$
$$C \longrightarrow B \longrightarrow D$$
$$D \longrightarrow B \longrightarrow E \longrightarrow C$$
$$E \longrightarrow B \longrightarrow A \longrightarrow C$$

Exercises

1. Complete the Kruskal's algorithm for the example in the text.
2. Another way to find a minimal spanning tree for a graph is to start with any vertex, then choose the cheapest edge coming out of that vertex. Next, considering the vertices already in the spanning tree, find the cheapest edge out of those vertices that doesn't introduce a cycle. Write an algorithm that captures this idea of starting with a vertex, rather than starting with the cheapest edge the way that

Kruskal's algorithm does. Note that there is no need to sort the edges in this approach. This approach is called Prim's algorithm.

3. Give an argument that Kruskal's algorithm does give a spanning tree with the least cost.

4. Give an argument that Prim's algorithm does give a spanning tree with the least cost. (See Exercise 2.)

5. Is it ever possible to get more than one cheapest cost spanning tree? If no, show why not. If yes, give an example to illustrate your claim.

6. Prove that the relation "is connected to by an edge" is an equivalence relation on the set of vertices for a given graph.

7. How many edges are there in a complete graph of n vertices?

8. Draw all possible connected graphs consisting of one vertex, with two vertices, with three vertices, with four vertices, with five vertices. Make a conjecture about numbers of each.

9. Prove or disprove: Every graph has a unique minimal spanning tree.

10. How many minimal spanning trees can a graph have?

11. Construct a graph where the vertices are you, your parents, and your grandparents. The edges will mean *born in the same state.*

12. Construct a graph where the vertices consist of the integers from 1 to 10 and each edge (x, y) means that x *divides* y.

13. Find a minimal spanning tree for the following graph:

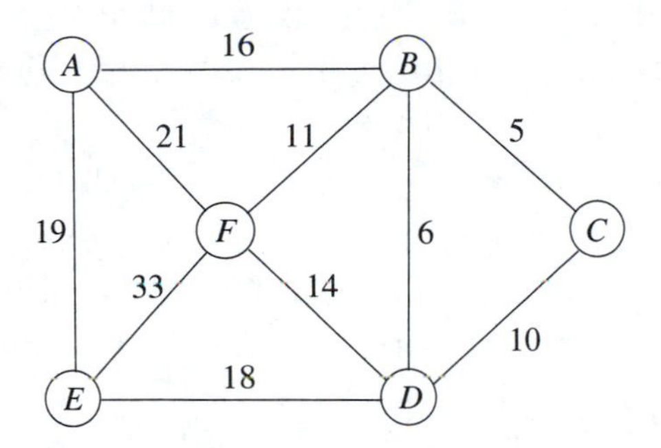

14. Does the following graph have an Euler circuit? An Euler tour? If one exists, find it.

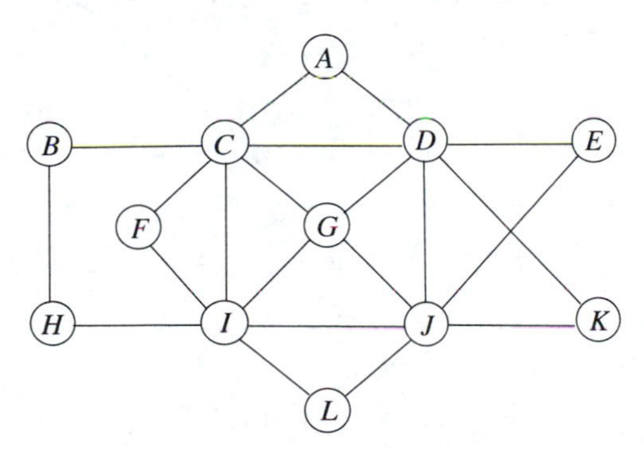

15. If land areas are represented as vertices and bridges as edges, the original Konigsburg bridge problem can not be represented as a graph, strictly speaking, since there would be more than one edge connecting two vertices. We call this sort of "graph" a *multigraph.* Draw the multigraph for the Konigsburg bridge problem.
    How would you store a multigraph as a two-dimensional array?

16. Suppose you are planning out what courses you need to take in your major. Represent possible ways you might fit them into your curriculum by drawing them on a graph. Your graph should represent each course as a node. When one course is required as a prerequisite to another, you should put an edge from the prerequisite to the course it is

a prerequisite for. There should be at least one course for which there is no prerequisite to get the graph started. Using the graph, determine at least three different sequences of courses you could take.

17. Exhibit a matrix to represent the graph Exercise 14. Your matrix will have 0's and 1's as entries, each entry representing whether or not there is an edge between the vertices whose numbers are the row and column numbers of the given entry.

18. Exhibit a matrix to represent the graph shown in the minimal spanning tree section of the chapter. Here the entries will not be limited to 0's and 1's. Instead, the entries will be the weights on the edges.

19. How many edges are in a graph of four vertices if every pair of vertices has exactly one edge joining that pair?

20. How many edges are in a graph of n vertices if every pair of vertices has exactly one edge joining that pair?

21. Suppose a connected graph has five vertices. What is the minimum number of edges?

22. Suppose a connected graph has n vertices. What is the minimum number of edges?

* 23. Suppose the National Football League has two conferences each with 13 teams. It was decided that each team would play a total of 14 games, 11 of which were to be with teams in their own conference and the other three with teams in the other conference. Why will this plan not work?

24. In a tree, because there are no cycles, there are some vertices that have edges with degree 1 (the vertex has an edge with only one other vertex). Those vertices are called leaves. Draw a tree with 15 vertices arranged so that every vertex in the tree (except for the *root*, which has two edges, and the leaves, which have one edge) has exactly three edges touching it. The tree you have drawn is called a *complete binary tree*. It is said to have *height* 3 because

the leaves are 3 edges away from the *root*.

25. How many leaves are in the complete binary tree of height 3?
26. Draw a complete binary tree of height 4. How many leaves? How many vertices?

* 27. How many leaves in the complete binary tree of height n? Prove your answer using induction. How many vertices?

Programming Problems

1. Write a program that reads in costs associated with the edges of a particular graph and prints out the edges (as ordered pairs of vertices) in a minimal spanning tree. *Hint*: It will be unnecessary to store the graph. You can test your program on the graph we used to illustrate this problem.
2. Write a program that determines whether or not a particular graph has an Euler circuit.
3. Write a program that prints out the edges of a weighted graph in order of their costs.
4. Write a program that, given two vertices in a directed graph, determines whether or not there is a path from the first vertex to the second. Will your program work for an undirected graph? Why or why not?
5. Write a program that stores an unweighted graph using a two dimensional matrix. Choose a vertex and find a path from that chosen vertex to each of the remaining vertices. Test your program on graphs of various sizes. Will your program always get the "best" path? What if you wanted paths that use the smallest number of edges?
6. Write a program that stores an unweighted graph using adjacency lists. Choose a vertex and find a path from that vertex to each of the other vertices in the graph. Keep track of how many edges your path requires. Test your

program on various graphs. Will your program always get the path with the least number of edges? If not, how can you modify it to do so?

7. A graph is connected if there is a path from any vertex to any other vertex. Write a program to determine whether or not a given graph is connected. Be sure to test it on graphs that are not connected as well as those that are.

Answers to Selected Odd-Numbered Exercises

Chapter 1

1. $\{E, L, P, H, A, N, T, S, Y, C, O\}$, $\{P, H, A, N, T\}$, $\{A, N, T\}$, $\{E, L, P, H, A, N, T, S, U, D\}$, $\{A, N, S, T\}$, $\{E, A, N, T\}$, $\{B, D, E, G, H, J, K, L, M, O, P, Q, R, U, V, W, X, Y, Z\}$, $\{B, G, J, K, M, Q, R, V, W, X, Z\}$
3. $\{2, 3, 5, 7, 11, 13, 17, 19, 23, 29, 31, 37, 41, 43, 47\}$
25. $\{(1, 2), (1, 3), (2, 2), (2, 3)\}$
27. No: $A \times B \neq B \times A$ if $A \neq B$
29. $\{-2, -3\}$
31. $\{x|x \text{ is an even integer}\}$
33. $A \cap B'$, $\emptyset$
35. $\{\emptyset, \{x, \{x\}\}, \{x\}, \{\{x\}\}\}$
39. A and B are disjoint.
41. $\{0\}$, $\{x : 0 \leq x \leq 1\}$
43. $\{(x, y) : y = x^2\}$
45. No
47. A
57. $A = B$

Chapter 2

5. $\lfloor \log_{10} n \rfloor + 1$
7. $\lfloor x + 0.5 \rfloor$
11. $\{a\}$
13. The equivalence relations are: $|a - b|$ is an integer, a has the same area as b, a and b have the same parents
21. Yes
23. $a \sim b$ iff "$a \leq b$ or $a - b$ is even"
25. $a \sim b$ iff $a < b$
27. (a) Yes, No; (b) Yes, No; (c) Yes, Yes; (d) No, Yes
29. $g \circ f(x) = -5x + 3$, $f \circ g(x) = 33 - 5x$
33. $f(x) = 5x$ is injective and surjective with inverse $g(x) = \frac{1}{5}x$, $f(x) = -1/(|x| + 1)$ is neither, $f(x) = x^2 + 1$ is neither, $f(x) = 2 - 3x$ is both withinverse $g(x) = (2 - x)/3$, $f(x) = x^{\frac{1}{2}}$ is injective but not surjective

Chapter 3

5. r
9. $B + C' + D$
13. 100101, 10001, 100000000000
15. 61, 34, 448
17. Move the binary point one place to the right to multiply by 2. To the left to divide.

19. 11101010.11

21. $(a+b)^n = a+b$

27. Theorem 1: For all a in B, $a * a = a$.
Theorem 2: For all a in B, $a + a = a$.

29. Duals do not exist.

51. AB', $ABC' + A'B'C$

53. $(\neg A \vee B) \vee (A \wedge B)$, $\neg(A \wedge (B \vee \neg C)) \vee C$

55. 2^4, 2^8, 2^{2^n}

Chapter 4

7. \$55,000

23. $(n^2 - 3n)/2$

27. 44

Chapter 5

3. 13, 6, 2

5. 31, 1000

7. 11, 13, 17, 19, 23, 29, 31, 37, 41, 43, 47, 53, 59, 61, 67, 71, 73, 79, 83, 89, 97

9. 4, 9, 866

11. 2, 8; none

13. The additive inverses for $0, 1, 2, 3, 4, 5, 6, 7, 8, 9$ in $\mathbb{Z}_{10}$ are $0, 9, 8, 7, 6, 5, 4, 3, 2, 1$. The additive inverses for $0, 1, 2, 3, 4, 5, 6, 7, 8$ in $\mathbb{Z}_9$ are $0, 8, 7, 6, 5, 4, 3, 2, 1$.

15. $2 \cdot 3 \cdot 5$, $2 \cdot 3^2 \cdot 5$, $2^2 \cdot 5^2$, 101, $2^3 \cdot 5^3$, $2 \cdot 3 \cdot 167$

25. 16, 1

27. life is a mystery, 12

29. $15c + 12$, $5c + 6$

Chapter 6

5. $d = 4$, $a = -3$, $b = 1$; $d = 1$, $a = -7$, $b = 13$

7. 82, 5

9. $2^{100} - 1$ sec $\approx 4 \times 10^{20}$ centuries

Chapter 7

1. $a_n = m^n d$

3. $T_n = n + 2$

5. $a_n = 4 \cdot 3^n + (3^n - 1)/2$

7. $a_n = c + bn$

9. $2^{n+1} + n(2^n - 1) - (2 + (n - 2)2^n) = 2^{n+2} - n - 2$

11. $b_n = 2 + 2\log_2 n$

13. $100 * (1.08)^n$

15. $a_0 = 1$, $a_n = (1.015)a_{n-1}$ (So $a_n = 1.015^n$) $a_{12} \approx 1.20$, $a_{60} \approx 2.44$, $a_{120} \approx 5.70$

Chapter 8

1. $\binom{12}{5} = 792$

3. $\binom{20}{12} = 125{,}970$,
$\binom{10}{6}\binom{10}{6} = 44{,}100$,
$\binom{10}{2}\binom{10}{10} + \binom{10}{4}\binom{10}{8} + \binom{10}{6}\binom{10}{6} + \binom{10}{8}\binom{10}{4} + \binom{10}{10}\binom{10}{2} = 63{,}090$,
$\binom{10}{10}\binom{10}{2} + \binom{10}{9}\binom{10}{3} + \binom{10}{8}\binom{10}{4} + \binom{10}{7}\binom{10}{5} = 40{,}935$
$\binom{10}{8}\binom{10}{4} + \binom{10}{9}\binom{10}{3} + \binom{10}{10}\binom{10}{2} = 10{,}695$

5. $4\binom{13}{5} = 5{,}148$, $4\binom{13}{5}/\binom{52}{5} \approx .00198$

7. $\binom{4}{3}\binom{4}{2} = 24$

9. $\binom{13}{2}\binom{4}{2}\binom{4}{2}\binom{44}{1} = 123{,}552$

11. $12! = 479{,}001{,}600$,
$4!8! = 967{,}680$,
$4!5!3! = 17{,}280$

13. $12!/(3!2!)$ vs $13!/(2!4!2!)$ or
39,916,800 vs 64,864,800
So PENNSYLVANIA

15. $\frac{4\cdot 6!}{2!2!} = 720$

17. $1 - 9^{10}/10^{10} \approx .65132$

19. $1 - (8\cdot 9^{n-1})/(9\cdot 10^{n-1})$,
limit as $n \to \infty$ is 1

21. $\binom{24}{3} = 2024$,
$\binom{6}{3} + \binom{6}{3} + \binom{6}{3} + \binom{6}{3} = 80$,
$(\binom{6}{3} + \binom{6}{3} + \binom{6}{3} + \binom{6}{3})/\binom{24}{3} \approx$ .0395

23. $11!/(2!2!2!) = 4{,}989{,}600$

25. $9 \cdot 10^6$

27. $7!/(3!3!) = 140$

29. $\binom{3}{1}\binom{3}{1}\binom{3}{1}/\binom{9}{3} = 9/28$

31. $\binom{20}{5}\binom{15}{5}\binom{10}{5}\binom{5}{5} = \frac{20!}{5!5!5!5!} =$ 11,732,745,024

33. $(1/2)^4 = 1/16$

35. $(1/2)^4 + (1/2)^4 = 1/8$

37. $\binom{5}{3}/\binom{11}{3} \approx .0601$

39. $9\cdot 9\cdot 8\cdot 7/(9\cdot 10^3) = .504$

41. $(2+2)/6^2 = 1/9$

43. $\binom{100}{20} =$
535,983,370,403,809,682,970

45. $\binom{5}{3} = 10$

47. $\binom{8}{4} = 70$, $\binom{8}{1} + \binom{8}{2} + \binom{8}{3} + \binom{8}{4} + \binom{8}{5} + \binom{8}{6} + \binom{8}{7} = 254$

49. $2^{10} = 1024$, $\binom{10}{3} = 120$,
$\binom{10}{3}/2^{10} \approx .11719$,
$\binom{10}{0} + \binom{10}{1} + \binom{10}{2} + \binom{10}{3} = 176$,
$(\binom{10}{0} + \binom{10}{1} + \binom{10}{2} + \binom{10}{3})/2^{10} \approx .1719$, $\binom{10}{5} = 252$,
$\binom{10}{5}/2^{10} \approx .2409$

Chapter 9

1. A is 4 by 4. B is 3 by 4. C is 4 by 4. D is 1 by 4. E is 3 by 4. F is 2 by 2.

3. AA, AC, BA, BC, CA, CC, DA, DC, EA, EC, FF

5. $AA = \begin{pmatrix} 53 & 76 & 38 & 36 \\ 31 & 58 & 25 & 27 \\ 31 & 25 & 15 & 11 \\ 30 & 30 & 20 & 11 \end{pmatrix}$

$AC = \begin{pmatrix} 27 & 39 & 47 & 59 \\ 15 & 27 & 34 & 63 \\ 11 & 6 & 12 & 20 \\ 15 & 6 & 8 & 8 \end{pmatrix}$

$BA = \begin{pmatrix} 16 & 23 & 10 & 15 \\ 28 & 45 & 25 & 15 \\ 8 & 1 & 0 & 4 \end{pmatrix}$

$BC = \begin{pmatrix} 7 & 25 & 28 & 17 \\ 18 & 5 & 7 & 31 \\ 0 & 9 & 12 & 3 \end{pmatrix}$

$CA = \begin{pmatrix} 24 & 26 & 15 & 12 \\ 50 & 12 & 8 & 12 \\ 35 & 65 & 24 & 43 \\ 25 & 10 & 3 & 14 \end{pmatrix}$

$CC = \begin{pmatrix} 11 & 12 & 15 & 14 \\ 6 & 10 & 26 & 26 \\ 16 & 72 & 81 & 64 \\ 2 & 27 & 36 & 17 \end{pmatrix}$

$DA = \begin{pmatrix} 11 & 13 & 7 & 5 \end{pmatrix}$

$DC = \begin{pmatrix} 5 & 2 & 4 & 12 \end{pmatrix}$

$EA = \begin{pmatrix} 63 & 88 & 46 & 40 \\ 14 & 3 & 0 & 10 \\ 29 & 36 & 18 & 22 \end{pmatrix}$

$EC = \begin{pmatrix} 33 & 41 & 49 & 61 \\ 0 & 25 & 30 & 3 \\ 13 & 35 & 39 & 19 \end{pmatrix}$

$FF = \begin{pmatrix} 4 & 0 \\ 9 & 1 \end{pmatrix}$

7. $\det A = 99$, $\det C = 324$, $\det F = 2$
9. a. $x_1 = 1$, $x_2 = 2$, $x_3 = 3$
   c. No solutions.
   e. $x_1 = \frac{23}{7} + x_3$, $x_2 = -\frac{1}{7} + x_3$
11. $\begin{pmatrix} 2 & 23 \\ 19 & 2 \end{pmatrix}$
15. NJ.K T
17. 3
19. 1

Chapter 10

1. Add (in this order) AG, DE, GI
5. Yes, if all edges have the same cost (for example).
7. $\binom{n}{2} = n(n-1)/2$
9. false
13. AB, BC, BF, BD, DE
21. 4
25. 8
27. 2^n leaves, $2^{n+1} - 1$ total vertices

Index